Astrophysics and Cosmology

A-Level

Astrophysics and Cosmology

Roger Muncaster
B.Sc Ph.D
Formerly Head of Physics
Bury Metropolitan College
of Further Education

Stanley Thornes (Publishers) Ltd.

First published in 1997 by
Stanley Thornes (Publishers) Ltd
Ellenborough House
Wellington Street
Cheltenham
Gloucestershire GL50 1YW
England

A catalogue record for this book is available from the British Library.

ISBN 0 7487 2865 1

97 98 99 00 / 10 9 8 7 6 5 4 3 2 1

The front cover shows a Hubble Space Telescope image of a small portion of the Cygnus Loop – the remnant of a star that exploded as a supernova about 15 000 years ago.

Typeset by Tech-Set, Gateshead, Tyne & Wear.
Printed and bound at Scotprint, Musselburgh.

Contents

Preface

This book is intended to cover the various Astronomy options of the current A-level and AS-level syllabuses. Specifically, it is aimed at Module PH04 of the NEAB syllabuses in A-level and AS-level Physics, Module PH3 (Topic 3C) of the London A-level Physics syllabus, Module 4837 of the Cambridge (modular) syllabuses in A-level Physics and A-level Science, and Option A of the Cambridge (linear) syllabuses in A-level and AS-level Physics.

Questions are included at relevant points in the text so that students can obtain an immediate test of their understanding of a topic. 'Consolidation' sections stress key points and in some cases present an overview of a topic in a manner which would not be possible in the main text. Definitions and fundamental points are highlighted – either by the use of screening or bold type. Questions, most of which are taken from past A-level papers, are included at the ends of chapters

Acknowledgements

I wish to express my gratitude to the publishers, and to John Hepburn and Margaret O'Gorman in particular, for their help throughout the preparation of the book.

I am indebted to the following examination boards for permission to use questions from their past examination papers:

Associated Examining Board [AEB]
University of Cambridge Local Examinations Syndicate [C], reproduced by permission of University of Cambridge Local Examinations Syndicate
Northern Examinations and Assessment Board (formerly the Joint Matriculation Board) [N]
Southern Universities' Joint Board [S]
Edexcel Foundation, London Examinations (formerly University of London Examinations and Assessment Council, and University of London School Examinations Board) [L]
University of Oxford Delegacy of Local Examinations [O]

Thanks are due to the following for providing photographs:

Ann Ronan Picture Library: p. 100(a)
NASA: front cover
Royal Astronomical Society: pp. 52, 77
Science Photo Library: pp. 9, 11(a), 12, 13, 36, 100(b), 105; 3 (Dr Fred Espenak); 4(a), 4(b) (NOAD); 4(a), 43 (Dr Rudolph Schild); 11(b) (Dr Jeremy Burgess); 44 (Lick Observatory); 45 (Robin Scagell); 56 (Hale Observatories); 83 (Martin Bond); 97 (Royal Greenwich Observatory).

R. MUNCASTER

Helmshore

1 INTRODUCTION

1.1 BACKGROUND

- **The Earth** is one of the nine known (major) planets that orbit the Sun.
- **The planets**, unlike stars and galaxies, have no light of their own – they merely reflect the light of the Sun.
- **The Sun** is a star, and like all stars, it generates its energy by thermonuclear fusion.
- All the stars visible with the naked eye as individual stars are members of our own Galaxy; they represent only a tiny fraction of the total number of stars in the Galaxy.
- Because light travels at a finite speed, we see a star not as it is now but as it was when the light left it – **the more distant the star, the farther back in time we are seeing**.

1.2 DISTANCE UNITS

The astronomical unit (AU) is the mean distance between the centre of the Earth and the centre of the Sun.

The light-year is the distance travelled by light in a vacuum in one year.

1 astronomical unit (AU) = 1.496×10^{11} m
1 light-year = 9.461×10^{15} m

1.3 THE SOLAR SYSTEM

The main constituents of the **Solar System** are the Sun, the nine planets and their satellites (or moons), the asteroids (or minor planets), and comets.

The Sun

The Sun is a fairly ordinary star consisting of about 70% hydrogen (by mass), 28% helium and 2% heavier elements – mainly carbon and oxygen. Energy generated by thermonuclear fusion reactions in the core of the Sun, at a temperature of about 1.5×10^7 K, travels outwards through the radiative and convective zones (see section 5.3) to the photosphere.

The photosphere is the visible surface of the Sun and is the lowest of the three main layers of the Sun's atmosphere. The photosphere is a few hundred kilometres in thickness. Most of the visible light that reaches the Earth comes from the photosphere. Its spectrum is that of a black body (see section 3.1) at a temperature of 5800 K – the temperature at the base of the photosphere and what is regarded as the surface temperature of the Sun. This continuous spectrum is crossed by hundreds of dark lines (**Fraunhofer lines**) caused by absorption in the upper, cooler layers of the photosphere (see section 3.7).

The layer immediately above the photosphere is the **chromosphere**. The chromosphere is a few thousand kilometres thick and its temperature ranges from about 4000 K at the top of the photosphere to about 7000 K where it meets the outer layer of the Sun's atmosphere, the **corona** – an extremely hot (up to 10^7 K), low-density gas that extends outwards from the chromosphere for millions of kilometres. The chromosphere and the corona are very much fainter than the photosphere and therefore are not normally visible. During total solar eclipses, however, the chromosphere can be seen as a pinkish ring of light surrounding the blacked out disc of the Sun; the corona appears as a white glow around that. The spectra of both the chromosphere and the corona are characterized by emission lines. The corona is so hot that it has very strong X-ray emission.

The Sun's corona as seen during the total solar eclipse of July 11, 1991

The planets

The planets (Table 1.1) move in elliptical orbits about the Sun. The plane in which the Earth's orbit lies is called the **ecliptic**. Only Pluto and Mercury have orbits which are significantly inclined to the ecliptic. All the planets orbit the Sun in the same direction as the Sun rotates about its own axis, i.e. anti-clockwise as seen from above the North Pole of the Earth

Table 1.1 The planets

Planet	*Mean distance from Sun in AU*	*Orbital period in years*	*Eccentricity (See Note (i))*	*Inclination to ecliptic in degrees*	*Mass of planet / Mass of Earth*	*Number of known satellites*
Mercury	0.387	0.241	0.206	7.0	0.055	None
Venus	0.723	0.615	0.007	3.4	0.815	None
Earth	1.000	1.000	0.017	Zero	1	1
Mars	1.524	1.881	0.093	1.9	0.107	2
Jupiter	5.203	11.86	0.048	1.3	317.9	16
Saturn	9.539	29.46	0.056	2.5	95.2	18
Uranus	19.19	84.01	0.046	0.8	14.53	15
Neptune	30.06	164.79	0.010	1.8	17.14	8
Pluto	39.53	248.54	0.248	17.1	0.0021	1

Notes

(i) The greater the **eccentricity** (e), the less circular the orbit.

$$b^2 = a^2 (1 - e^2)$$

where a is the semimajor axis and b is the semiminor axis (see Fig. 1.1). A circular orbit would have an eccentricity of zero.

(ii) Pluto has a highly eccentric orbit and is sometimes closer to the Sun than Neptune (Fig. 1.2).

(iii) Mass of Earth $= 5.976 \times 10^{24}$ kg

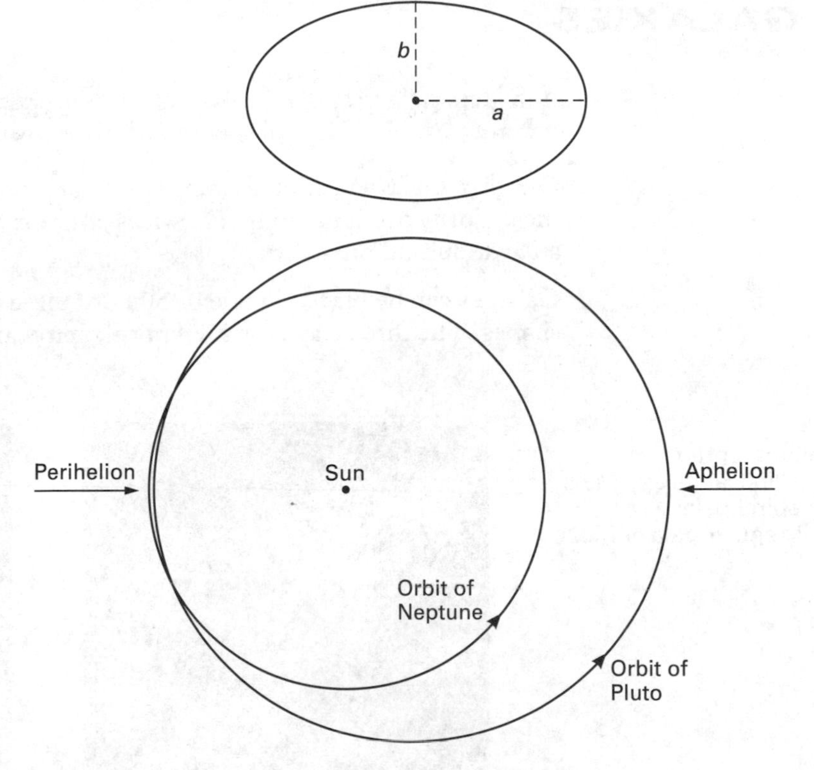

Fig. 1.1
Eccentricity of an ellipse

Fig. 1.2
Orbits of Neptune and Pluto

The Asteroids

The majority of the asteroids orbit the Sun in the so-called **asteroid belt** between the orbits of Mars and Jupiter. They range in size from **Ceres** (the largest, with a diameter of 913 km) down to particles no bigger than a grain of sand. They were once thought to be the remnants of a planet that had disintegrated. This is no longer believed to be the case – they almost certainly originate from the formation of the solar system, and have been prevented from combining to form a planet by the powerful gravitational effect of Jupiter. In any case, the total mass of the asteroids is only about one tenth that of the Moon.

Comets

Almost all of the mass of a comet is in its **nucleus**, which is typically only a few cubic kilometres in volume. The nucleus is covered by a thin crust and is composed of frozen material (mainly frozen water) in which dust and rock fragments are embedded. As a comet approaches the Sun, the frozen material in the nucleus starts to evaporate. Gas and dust break through the crust and create a large, but extremely tenuous, **coma** that shines by the reflected light of the Sun. The coma may be as much as a million kilometres in diameter.

Some (by no means all) comets develop one or more tails. The tails always point away from the Sun and are of two types – gas tails and dust tails. **Gas (or ion) tails** are composed of ionized molecules driven out of the coma by the **solar wind** (the stream of charged particles continuously ejected by the Sun). **Dust tails** consist of particles of dust blown from the coma by radiation (photon) pressure from the Sun. Gas tails are straight and may be over 100 million kilometres long; dust tails are shorter, broader and curved.

1.4 GALAXIES

A galaxy is a huge assembly of stars, gas and dust held together by gravity.

Our own Galaxy* contains about 10^{11} stars. Some galaxies exist in isolation, but the majority occur in groups known as **clusters** with anything from a few dozen to a few thousand members.

Galaxies can be placed in one or other of three broad classes on the basis of their shapes. The three classes are: elliptical, spiral and irregular (Fig. 1.3).

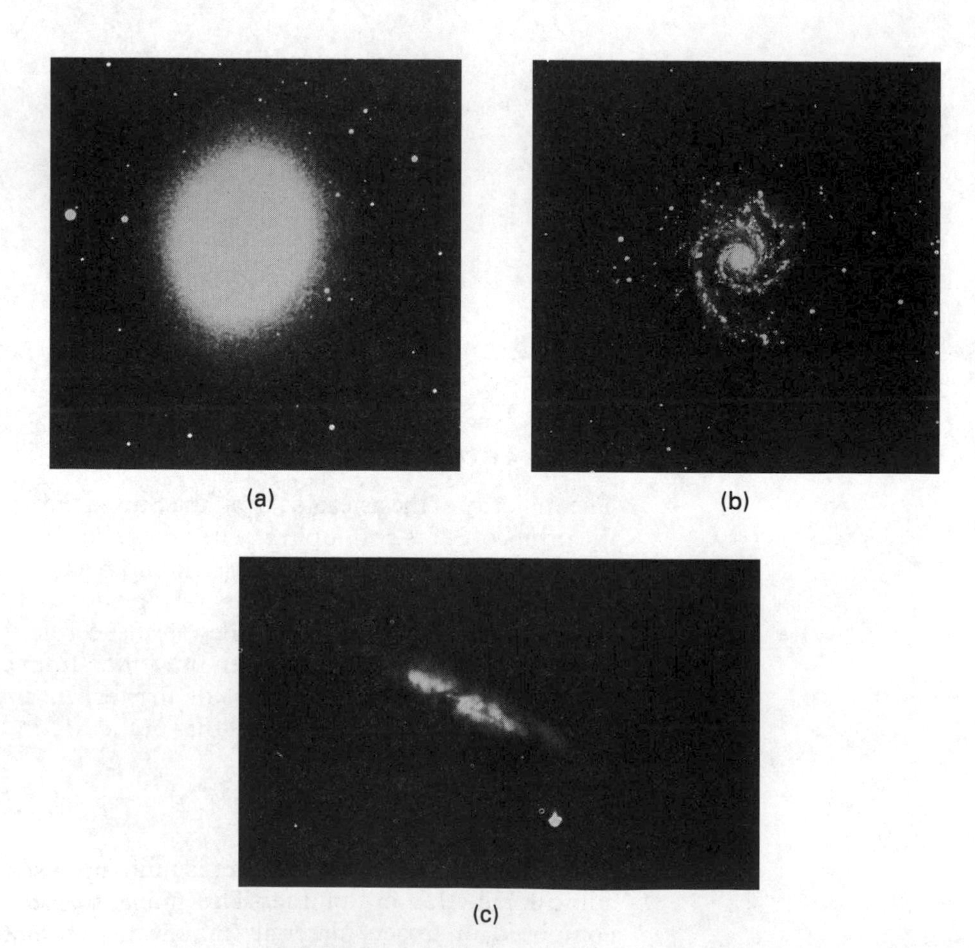

Fig. 1.3
Optical photographs of
(a) the elliptical galaxy M32,
(b) the spiral galaxy M74,
(c) the irregular galaxy M82

Elliptical galaxies are ellipsoidal; their shapes range from that of an almost perfect sphere to that of a rugby ball. Ellipticals consist mainly of old stars – they

*Note the use of a capital G when referring to our own Galaxy.

have very few newly formed stars and very little gas or dust. They come in a wider range of sizes than either spirals or irregulars – the **giant ellipticals** contain up to 10^{13} stars, the **dwarf ellipticals** have as few as 10^6.

Spiral galaxies are disc-shaped with spiral arms curving out from a **central bulge** that surrounds the **nucleus**. The stars in the central bulge, most of which are old stars, are more highly concentrated than elsewhere. The arms are composed mainly of bright, young stars and of gas and dust from which new stars are forming. Spiral galaxies rotate in the plane of the disc.

Irregular galaxies have no particular shape or structure and tend to be smaller than an average-sized elliptical or spiral galaxy. They often contain huge amounts of gas and dust from which new stars are forming.

1.5 OUR GALAXY

Our own Galaxy (sometimes called the **Milky Way Galaxy**) is a fairly typical spiral galaxy (section 1.4) containing about 10^{11} stars. The galactic disc has a diameter of about 100 000 light-years and is about 2000 light-years in thickness (Fig. 1.4). The disc has a high concentration of gas and dust contained in a thinner disc (the gas disc) about 500 light-years thick. The Sun is situated towards the edge of one of the spiral arms, some 28 000 light-years from the centre and about 25 light-years above the Galactic plane. The total mass of the Galaxy is estimated to be of the order of 10^{42} kg, with an average density of the order of 1×10^{-20} kg m^{-3}.

Fig. 1.4
The Milky Way Galaxy

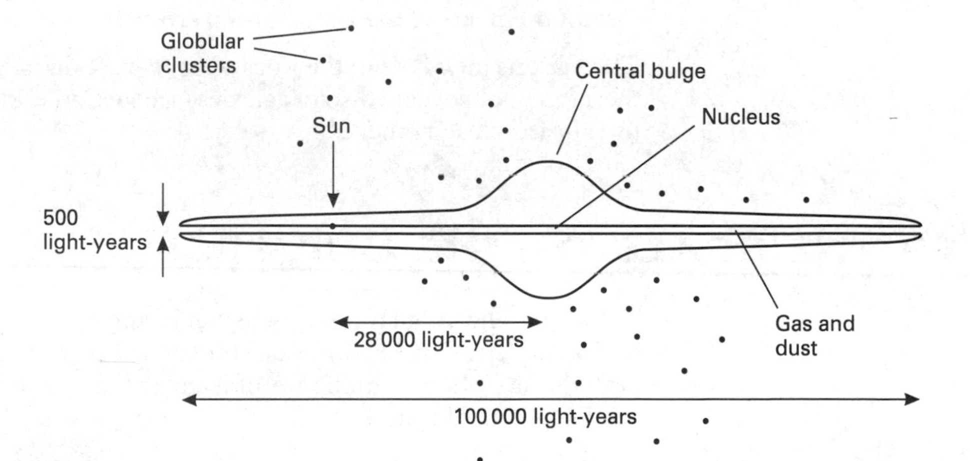

The rotation of the Galaxy

The Galaxy rotates about its centre, but not like a solid body. **The orbital periods of the stars in the Galactic disc increase with distance from the centre**. The same is true of the orbits of the planets about the Sun, but the exact relationship between orbital period and orbital radius is not the same. The planets move in accordance with Kepler's third law (section 2.3); the stars do not. We shall explain why in section 2.9. The Sun has an orbital speed of about 240 km s^{-1} and takes about 220 million years to complete its orbit.

The Galactic halo

The Galactic disc and central bulge are surrounded by the much more tenuous **Galactic halo** – a nearly spherical distribution of very old stars and about 140 globular clusters. **Globular clusters** are densely packed groups of ancient stars. They are approximately spherical, each containing between 50 thousand and 50 million stars. They are thought to be the oldest objects in the Galaxy, having formed over 10 000 million years ago when the rest of the Galaxy was still in the process of condensing from a huge cloud of gas and before it took on the disc-like shape it has now. The globular clusters rotate around the centre of the Galaxy, most of them in orbits that are highly inclined to the Galactic plane.

The Milky Way

The Milky Way is the faint, apparently continuous, band of light that (under suitable conditions) can be seen stretching across the sky from horizon to horizon. Telescopes reveal that it is actually an enormous number of faint stars packed closely together. The stars are the stars of the Galactic disc as seen from inside one of the spiral arms. When we look at the Milky Way we are looking along the plane of the Galactic disc.

1.6 THE LOCAL GROUP

Our own Galaxy is one of a cluster of about 30 galaxies known as the **Local Group**. The Andromeda galaxy (M31) and our own Galaxy (both spirals) are the two largest members. The nearest members of the Local Group are the Large and the Small Magellanic Clouds – two small irregular galaxies visible with the naked eye in the Southern Hemisphere. They are about 200 000 light-years away; the most distant members of the Local Group are over 2 million light-years away.

The various members of the Local Group are bound together gravitationally and therefore, unlike more distant galaxies (see section 7.1), they are not receding from us (nor from each other).

1.7 THE SCALE OF THE UNIVERSE

Tables 1.2 and 1.3 may help the reader to appreciate the scale and structure of the Universe. The Universe is vast, but its average density is remarkably small – perhaps as little as a millionth of a millionth of a millionth of a millionth of a millionth that of water!

Table 1.2
Some astronomical distances

	Distance in kilometres	*Distance in light-years*
Radius of Earth	6.4×10^3	
Earth to Moon	3.8×10^5	
Radius of Sun	7.0×10^5	
Earth to Sun	1.5×10^8	
Sun to Pluto	5.9×10^9	
Sun to nearest star	4.0×10^{13}	4.2
Sun to centre of Galaxy	2.8×10^{17}	3.0×10^4
Radius of Galactic disc	4.7×10^{17}	5.0×10^4
Sun to nearest galaxy	1.6×10^{18}	1.7×10^5
Most distant naked-eye object	2.1×10^{19}	2.2×10^6
Radius of known Universe	$\sim 1 \times 10^{23}$	$\sim 1 \times 10^{10}$

Table 1.3
Some astronomical masses

Object	*Mass / kg*
Moon	7.4×10^{22}
Earth	6.0×10^{24}
Jupiter	1.9×10^{27}
Sun	2.0×10^{30}
Typical galaxy	$\sim 1 \times 10^{42}$

2

THE HELIOCENTRIC UNIVERSE

2.1 THE PTOLEMAIC MODEL

Until the mid-sixteenth century the most widely accepted view of the Universe was the **geocentric (Earth-centred)** model that had been developed by Ptolemy (Claudius Ptolemaeus) around AD 150, and which had existed in one form or another from before the time of Aristotle (around 350 BC).

According to the **Ptolemaic model**, the Earth was motionless at the centre of the Universe and the Sun, Moon, planets and stars moved around the Earth.

The Ptolemaic model was based on the Greek belief that the perfect form of motion was a circle – the various bodies were therefore supposed to move in circles. Planets normally appear to move from west to east relative to the background of distant stars. There are occasions, however, when a planet will appear to stop and then start to move 'backwards' (i.e. from east to west) for several weeks. In order to explain this so-called **retrograde motion**, the planets were supposed to move in small circles (**epicycles**) whose centres moved around the Earth on larger circles (**deferents**) (Fig. 2.1). The model seemed to work at first, but as astronomical observations became ever more accurate, more and more adjustments had to be made to the basic model to make it fit the observations. The more complex it became, the less convincing it was, and it was this that led Copernicus to reject it completely.

Fig. 2.1
Explanation of retrograde motion according to the geocentrc theory

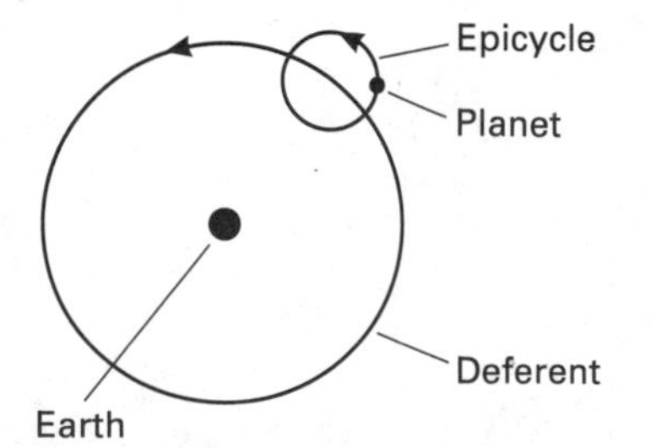

The planet spends most of its time moving anti-clockwise about the Earth, but because its speed along the epicycle is assumed to be greater than the speed of the epicycle along the deferent, the planet moves clockwise when it is on that part of the epicycle that is closest to the Earth.

2.2 THE COPERNICAN MODEL

In 1543 a Polish clergyman and astronomer, Nicolaus Copernicus, published a book (*De Revolutionibus Orbium Coelestium* – 'On the Revolution of the Celestial Spheres') in which he proposed that the Sun, and not the Earth, was at the centre of the Universe. This **heliocentric (Sun-centred)** theory had first been suggested by Aristarchus of Samos around 300 BC. Copernicus felt that it provided a very much simpler explanation of planetary motions than the Ptolemaic system.

Nicolaus Copernicus (1473–1543)

According to the **Copernican model** the Earth and the other planets orbit the Sun, and the Earth turns on its axis once every 24 hours.

The Copernican model provides a simple explanation for the retrograde motions of the planets and for their apparent changes in brightness. The brightness of Mars, for example, changes as its distance from the Earth changes, and this is simply the result of the Earth and Mars having different angular velocities about the Sun (Fig. 2.2). The retrograde motion is merely a natural consequence of the changing relative positions of the Earth and Mars. Despite this success, it was received with extreme hostility when it was first announced. Some of the objections are listed below.

(i) The fact that objects fall vertically was taken to be strong evidence that the Earth could not be moving.

(ii) It was argued that there would be an ever present wind due to the motion of the Earth through the air. Copernicus explained the absence of such a wind by claiming that the Earth dragged the atmosphere along with it.

(iii) The objectors claimed that if the Earth were moving around the Sun, then the positions of the stars should appear to change relative to each other during the course of the Earth's orbit. No such **stellar parallax** had ever been observed. Copernicus argued that this was simply because the stars were so far away that the parallax was too small to be detected. This argument was eventually proved to be correct but it commanded very little support at the time.

(iv) The basic model (planets moving on circular orbits) was not able to predict the positions of the planets as accurately as the Ptolemaic system and Copernicus was eventually forced to use epicycles to improve the agreement.

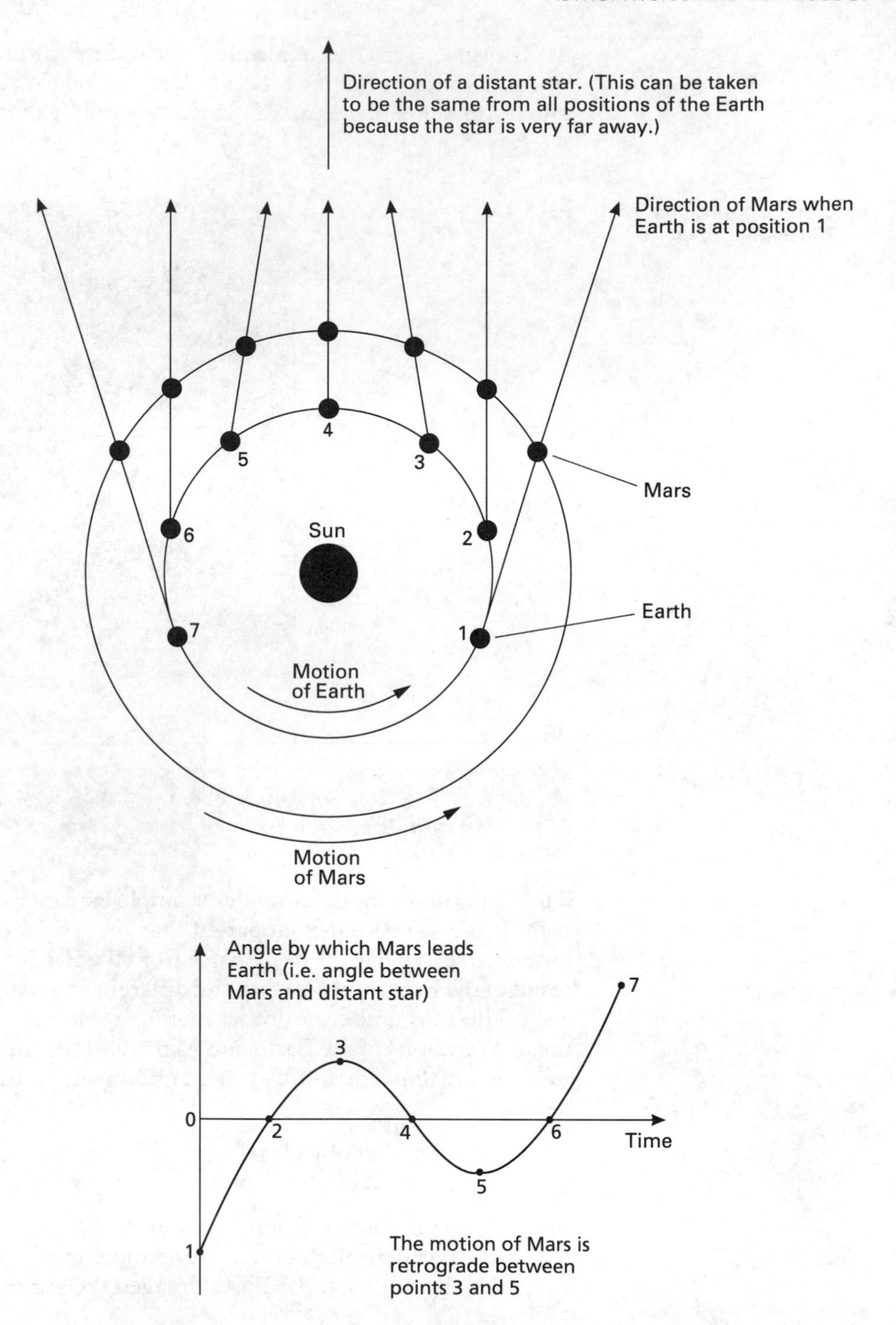

Fig. 2.2
Diagram to show how the Copernican model accounts for the retrograde motion of Mars

2.3 KEPLER'S LAWS

For about 20 years before his death in 1601, the Danish astronomer Tycho Brahe made precise measurements of the positions of the various bodies in the Solar System. His assistant, the German astronomer and mathematician, Johannes Kepler (1571–1630), made a detailed analysis of Brahe's data and was able to show that **the heliocentric system fits the observations if the planets move on elliptical, rather than circular, orbits.** Kepler published this discovery in 1609, and by 1619 he had announced a total of three laws which describe planetary motion.

Tycho Brahe (1546–1601)
Johannes Kepler (1571–1630)

(a)

(b)

1 The orbit of each planet is an ellipse which has the Sun at one of its foci.

2 Each planet moves in such a way that the (imaginary) line joining it to the Sun sweeps out equal areas in equal times.

3 The squares of the periods of revolution of the planets about the Sun are proportional to the cubes of their mean distances from it.

Fig. 2.3 illustrates law 2 but gives a very much exaggerated idea of the eccentricity of the planetary orbits – the ellipses along which the planets move are very nearly circular.

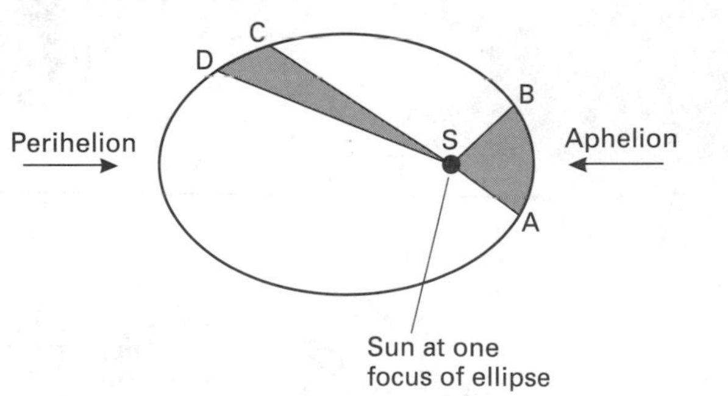

Fig. 2.3
Illustration of Kepler's second law

The accuracy and elegant simplicity with which Kepler's laws describe planetary orbits did much to establish the validity of the heliocentric theory.

QUESTIONS 2A

1. Plot a suitable graph using data from Table 1.1 to confirm the validity of Kepler's third law.

2. The average orbital radii about the Sun of the Earth and of Mars are 1.5×10^{11} m and 2.3×10^{11} m respectively. How many (Earth) years does it take Mars to complete its orbit?

2.4 GALILEO GALILEI

Galileo Galilei (1564–1642) was a contemporary of Kepler and, like Kepler, was a confirmed supporter of the heliocentric theory. In July 1609, Galileo heard of the invention of the telescope and quickly saw its potential as a means of studying the heavens. By December 1610 he had made a number of discoveries that gave strong support to the Copernican system.

Galileo Galilei (1564–1642)

(i) He observed that Venus goes through phases and that it appears larger when it is a crescent than when it is nearly full (Fig. 2.4). The Copernican theory provides a simple explanation for this – the geocentric theory cannot.

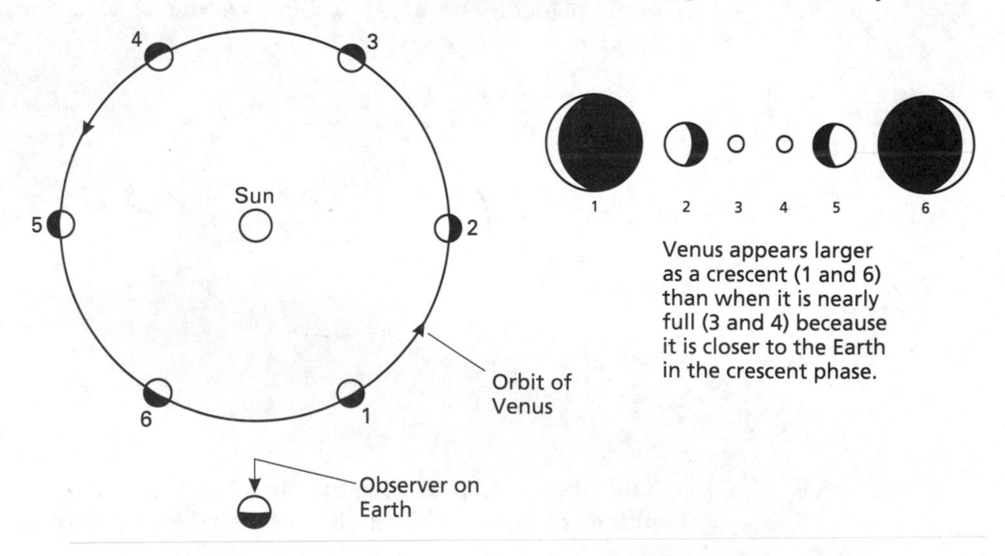

Fig. 2.4 Diagram to explain the relationship between the different phases of Venus and its apparent size

(ii) He discovered four satellites in orbit around Jupiter – proof that some heavenly bodies, at least, did not orbit the Earth.

(iii) The stars still appeared as tiny points of light when seen through a telescope, thus supporting Copernicus' contention that they are very far away.

(iv) He found that there were mountains and craters on the Moon – it was therefore not the perfect sphere that had hitherto been supposed. The geocentric lobby claimed that the Earth's unique position at the centre of the Universe distinguished it from the rest of the Universe and that it was the only imperfect body in the Universe.

Galileo's endorsement of the heliocentric theory brought him into conflict with the Vatican, and he spent the final eight years of his life under house arrest as a result. It was during this period that he did much of the work that was to provide the foundation for Newton's laws of motion. It is worth pointing out that these laws explain why objects fall vertically even though the Earth is moving and they therefore invalidate another of the objections to the Copernican theory.

2.5 NEWTON'S LAW OF UNIVERSAL GRAVITATION

About fifty years after Kepler's laws had been announced, Isaac Newton showed that any body which moves about the Sun in accordance with Kepler's second law must be acted on by a force which is directed towards the Sun. He was able to show that if this force is inversely proportional to the square of the distance of the body from the Sun, then the body must move along a path which is a conic section (i.e. elliptical, circular, parabolic or hyperbolic). Newton then showed that when the path is elliptical or circular the period of revolution is given by Kepler's third law. Thus, a centrally directed inverse square law of attraction is consistent with all three of Kepler's laws. Newton proposed that the planets are held in their orbits by just such a force. He further proposed that it is the same type of force which maintains the Moon in its orbit about the Earth, and which the Earth exerts on a body when it causes it to fall to the ground. Extending these ideas, Newton proposed that every body in the Universe attracts every other with a force which is inversely proportional to the square of their separation. His next step was to turn his attention to the masses of the bodies involved.

Isaac Newton
(1642–1727)

According to Newton's third law, if the Earth exerts a force on a body, then that same body must exert a force of equal magnitude on the Earth. Newton knew that the force exerted on a body by the Earth is proportional to the mass of the body. He saw no reason why the body should behave any differently from the Earth, in which case the force exerted on the Earth by the body must be proportional to the mass of the Earth. Since the two forces are equal, a change in one must be accompanied by an equal change in the other. It follows that each force must be proportional to the product of the Earth's mass and the mass of the body.

The ideas of the last two paragraphs are summarized in **Newton's law of universal gravitation.**

Every particle in the Universe attracts every other with a force which is proportional to the product of their masses and inversely proportional to the square of their separation.

Thus

$$F = G\frac{m_1 m_2}{r^2} \quad [2.1]$$

where

F = the gravitational force of attraction between two particles whose masses are m_1 and m_2, and which are a distance r apart

G = a constant of proportionality known as the **universal gravitational constant** ($= 6.67 \times 10^{-11}\,\mathrm{N\,m^2\,kg^{-2}}$).

Note Equation [2.1] is concerned with particles (i.e. point masses) but, in the circumstances listed below, it can also be used for bodies of masses m_1 and m_2 whose centres are a distance r apart.

(i) It is valid for two bodies of any size provided that they each have spherical symmetry. (The Sun and the Earth is a good approximation.)

(ii) It is a good approximation when one body has spherical symmetry and the other is small compared with the separation of their centres (e.g. the Earth and a brick).

(iii) It is a good approximation when neither body has spherical symmetry, but where both are small compared with the separation of their centres (e.g. two bricks a few metres apart).

2.6 TO SHOW THAT KEPLER'S THIRD LAW IS CONSISTENT WITH $F = G\frac{m_1 m_2}{r^2}$

Consider a planet of mass m moving about the Sun in a circular* orbit of radius r. Suppose that the mass of the Sun is m_S and that the angular velocity of the planet is ω (Fig. 2.5).

The force F which provides the centripetal acceleration $\omega^2 r$ is given by Newton's second law as

$$F = m\omega^2 r \quad [2.2]$$

By Newton's law of universal gravitation (equation [2.1])

$$F = G\frac{m m_S}{r^2}$$

*The mathematics required to treat the general case of an elliptical orbit is beyond the scope of this book but leads to the same result.

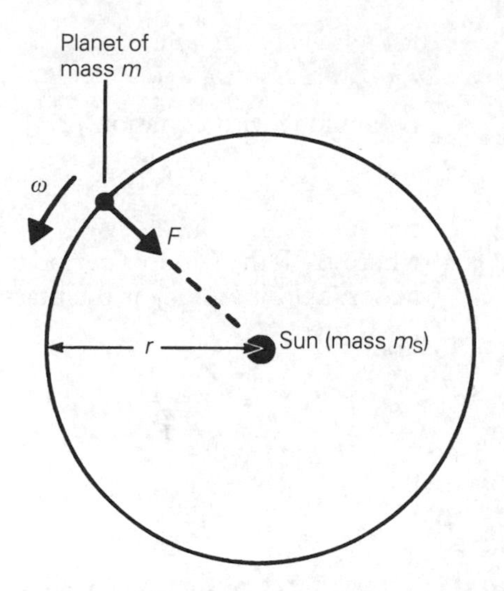

Fig. 2.5
Force on planet in circular motion around the Sun

Therefore, from equation [2.2]

$$G\frac{mm_S}{r^2} = m\omega^2 r$$

But $\omega = 2\pi/T$, where T is the period of revolution of the planet, and therefore

$$G\frac{mm_S}{r^2} = m\frac{4\pi^2}{T^2}r$$

i.e.

$$T^2 = \frac{4\pi^2}{Gm_S}r^3 \qquad [2.3]$$

Since G, m_S and π have the same values no matter which planet is being considered,

$$T^2 \propto r^3$$

This is Kepler's third law; it has been derived on the basis of Newton's law of universal gravitation and therefore the two laws are consistent.

Note Newton was not able to calculate the periods of the planetary orbits on the basis of equation [2.3] because he did not know the values of G and m_S.

2.7 NEWTON'S TEST OF THE LAW OF GRAVITATION

The last two sections make it clear that Newton's law of gravitation is consistent with Kepler's laws of planetary motion. However, the forces which hold the planets in their orbits are due, in every case, to the Sun. In order to show that gravitational attraction is universal, Newton needed to test it in circumstances that did not involve the Sun. The obvious test was to apply his ideas to the Earth–Moon system.

If a body of mass m is at the surface of the Earth, the force acting on the body is its weight, mg. This same force is given by the law of gravitation as Gmm_E/r_E^2 where m_E and r_E are respectively the mass and the radius of the Earth. Therefore

$$G\frac{mm_E}{r_E^2} = mg$$

i.e. $$G = \frac{gr_E^2}{m_E} \qquad [2.4]$$

By analogy with equation [2.3]

$$T^2 = \frac{4\pi^2 r_M^3}{Gm_E} \qquad [2.5]$$

where T is the Moon's period of revolution about the Earth, r_M is the radius of the Moon's orbit and m_E is the mass of the Earth. Substituting for G from equation [2.4] gives

$$T^2 = \frac{4\pi^2 r_M^3}{gr_E^2}$$

i.e. $$g = \frac{4\pi^2 r_M^3}{T^2 r_E^2} \qquad [2.6]$$

The value of r_E that was available to Newton was poor by present day standards. Even so, equation [2.6] gave a value for g that was sufficiently close to the accepted value for Newton to conclude that the Earth exerted the same type of force on the Moon as the Sun did on the planets.

2.8 THE MASSES OF THE SUN AND ITS PLANETS

In 1798, a hundred and twenty-one years after Newton had proposed the law of universal gravitation, Henry Cavendish made the first reliable measurement of G. Once this had been done, it was possible to obtain a value for the mass of the Sun on the basis of equation [2.3] and the radius and period of any planetary orbit. The mass of any planet that has a moon can be found by using an analogous equation and the radius and period of the moon's orbit about the planet. The mass of the Earth, for example, can be found from equation [2.5].

Note Cavendish used two lead spheres attached (one) to each end of a horizontal beam suspended by a silvered copper wire. When two larger spheres were brought up to the smaller ones, they deflected under the action of the gravitational force and twisted the suspension. The wire had been calibrated previously so that the strength of the force could be determined by measuring the angle through which the wire was twisted. An improved version of the experiment was performed by Boys in 1895.

QUESTIONS 2B

1. Find the gravitational force of attraction between two 10 kg particles which are 5.0 cm apart. ($G = 6.7 \times 10^{-11}\,\mathrm{N\,m^{-2}\,kg^{-2}}$.)

2. Calculate the mass of the Earth by considering the force it exerts on a particle of mass m at its surface. (Radius of Earth $= 6.4 \times 10^3$ km, $g = 9.8\,\mathrm{m\,s^{-2}}$, $G = 6.7 \times 10^{-11}\,\mathrm{N\,m^{-2}\,kg^{-2}}$.)

3. Io, the innermost of the 4 Galilean satellites of Jupiter, has an orbital period of 1.77 days and a mean orbital radius of 4.22×10^5 km. Calculate the mass of Jupiter. ($G = 6.67 \times 10^{-11}\,\mathrm{N\,m^{2}\,kg^{-2)}}$.)

2.9 THE MASSES OF GALAXIES

Consider a star of mass m in a circular orbit about the centre of its galaxy and at a distance r from the centre. The gravitational force of attraction that provides the star's centripetal acceleration (v^2/r) is due only to that portion of the galaxy that lies inside the orbit of the star, and therefore we may put

$$G\frac{mM}{r^2} = \frac{mv^2}{r}$$

where M is the mass of the galaxy that is inside the star's orbit and v is the orbital speed of the star. Rearranging gives

$$M = \frac{v^2 r}{G} \qquad [2.7]$$

Equation [2.7] can be used to estimate the mass of the Galaxy that is inside the Sun's orbit, for example. Putting $v = 240\,\text{km s}^{-1}$ ($= 2.4 \times 10^5\,\text{m s}^{-1}$) and $r = 28\,000$ light-years ($= 2.6 \times 10^{20}\,\text{m}$) gives $M = 2.2 \times 10^{41}\,\text{kg}$.

Notes

(i) Radio telescopes have been used to determine the orbital speeds of the clouds of gas that exist far from the centres of galaxies. These studies indicate that as much as 90% of the mass of a spiral galaxy might lie outside the visible edge of its disc. No one has ever detected any radiation coming from this so-called **dark matter** and its nature is the subject of considerable speculation. (The mass of the gas that betrays the presence of the dark matter is far too small to account for the 'missing' mass.)

(ii) The planets obey Kepler's third law because they are all acted on by the gravitational attraction of the same mass – the mass of the Sun. As we have just seen, the various stars in a galaxy are each under the influence of a different mass and therefore do not obey Kepler's third law.

2.10 THE MASSES OF BINARY STARS

A binary star consists of two stars in orbit about their common centre of mass. Consider two such stars of masses m_1 and m_2 whose separation is r and which are in circular orbits of radii r_1 and r_2 respectively about their centre of mass (Fig. 2.6).

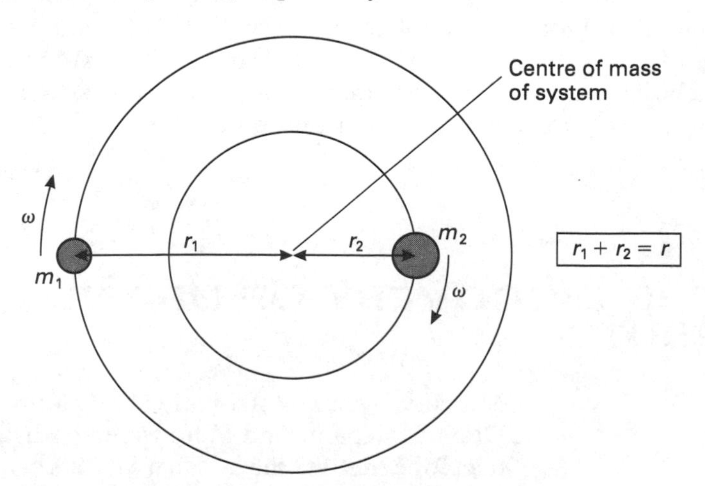

Fig. 2.6
A binary star

It follows from the definition of centre of mass that

$$m_1 r_1 = m_2 r_2$$
$$= m_2(r - r_1)$$

i.e.
$$r_1 = \frac{m_2 r}{m_1 + m_2} \qquad [2.8]$$

If the angular velocity of the stars is ω, it follows from the law of gravitation that for the motion of the star of mass m_1

$$G\frac{m_1 m_2}{r^2} = m_1 \omega^2 r_1$$

Therefore by equation [2.8]

$$G\frac{m_1 m_2}{r^2} = \frac{m_1 \omega^2 m_2 r}{m_1 + m_2}$$

$$\therefore \quad m_1 + m_2 = \frac{\omega^2 r^3}{G}$$

If the orbital period is T, $\omega = 2\pi/T$ and therefore

$$m_1 + m_2 = \frac{4\pi^2 r^3}{GT^2} \qquad [2.9]$$

Equation [2.9] allows us to find the combined mass of the stars providing we know their period and their separation. Once we have found the combined mass, we can find the individual masses providing we also know the distance of either star from the centre of mass.

Note In deriving equation [2.3] we assumed that the planet orbited the Sun rather than their common centre of mass. Since the mass of the Sun is over 1000 times that of even the most massive of the planets (Jupiter), the common centre of mass is very close to the centre of the Sun and this simplified approach leads to very little error in the case of planetary orbits.

QUESTIONS 2C

1. The stars in a binary system have an orbital period of 1.4×10^8 s and are 4.5×10^8 km apart. The lighter of the two stars is at a distance of 3.6×10^8 km from their common centre of mass.

   Find
   (a) the combined mass of the stars,
   (b) the mass of the heavier star.
   ($G = 6.7 \times 10^{-11}\,\mathrm{N\,m^{-2}\,kg^{-2}}$.)

2.11 THE DISCOVERY OF URANUS, NEPTUNE AND PLUTO

Mercury, Venus, Mars, Jupiter and Saturn have been known since ancient times; Uranus, Neptune and Pluto were not discovered until after the invention of the telescope (despite the fact that Uranus can be seen with the naked eye).

Uranus was discovered by William Herschel in 1781 whilst conducting a detailed telescopic survey of the sky. (Strictly, Herschel was the first to recognize that Uranus is a planet – it had been seen many times before but had always been taken to be an ordinary star.)

The planets feel the gravitational attraction of each other as well as that of the Sun. This leads to **perturbations** (irregularities) in their orbits, and these can be calculated using Newton's law of gravitation. Some years after Uranus had been discovered, it became clear that there were discrepancies in its orbit that could not be accounted for by the effects of the known planets. It was unlikely that there was anything wrong with the law of gravitation, for it correctly predicted the orbits of the other planets, and therefore it seemed likely that there was an unknown planet affecting the orbit of Uranus.

A French astronomer, Urbain Le Verrier, calculated where this unknown planet might be, and in September 1846, Johann Galle, of the Berlin Observatory, found **Neptune** within 1° of where Le Verrier had said it would be.

Note An English astronomer, John Couch Adams, had calculated the position of Neptune independently of Le Verrier, and about 12 months before him, but failed to convince anyone to mount a search for it. After much wrangling, Adams and Le Verrier were jointly credited with having made the prediction that led to the discovery.

There are irregularities in the motion of Uranus that cannot be attributed to the presence of Neptune. Two American astronomers, Percival Lowell and William Henry Pickering, independently attributed this to an unknown planet beyond the orbit of Neptune and calculated its expected position. (Such a planet would perturb the orbit of Neptune more than that of Uranus – the calculations were based on irregularities in the motion of Uranus because Neptune's orbit was not sufficiently well known to allow a reliable calculation.) Their calculations eventually led to the discovery of **Pluto** by Clive Tombaugh in 1930.

CONSOLIDATION

Until the mid-sixteenth century most astronomers believed that the Earth was motionless at the centre of the Universe.

According to the Copernican model the Earth and the other planets orbit the Sun. This provides a much simpler explanation of the retrograde motions of the planets and their apparent changes in brightness.

Galileo's discoveries with the telescope and Kepler's laws of planetary motion did much to establish the validity of Copernicus' heliocentric model.

Newton's Law of Universal Gravitation

$$F = G\frac{m_1 m_2}{r^2}$$

The law of gravitation accounts for all three of Kepler's laws.

Applying the law of gravitation to perturbations in the orbit of Uranus led to the discoveries of Neptune and Pluto.

QUESTIONS ON CHAPTER 2

1. **(a)** Outline the contributions made by Kepler and by Galileo that helped to establish Copernicus' heliocentric theory of the Universe.
(b) Why would it be better to refer to Copernicus' theory as the heliocentric theory of the Solar System?

2. **(a)** Explain why a planet nearer to the Sun than the Earth (e.g. Venus) can be seen through a small telescope to exhibit phases like the Moon, changing from a crescent shape to a full circular disc and back.
Suggest a reason why the planet Mars does not exhibit phases.
(b) Draw a diagram showing the relative positions of the Sun, Earth and Venus at a time when exactly half of the illuminated surface of Venus is visible from the Earth. Show on your drawing the paths of light reaching the observer. [N*]

3. Kepler's third law of planetary motion, as simplified by taking the orbits to be circles round the Sun, states that if r denotes the radius of the orbit of a particular planet and T denotes the period in which that planet describes its orbit, then r^3/T^2 has the same value for all the planets.
The orbits of the Earth and of Jupiter are very nearly circular with radii of 150×10^9 m and 778×10^9 m respectively, while Jupiter's period round the Sun is 11.8 years.
(a) Show that these figures are consistent with Kepler's third law.
(b) Taking the value of the gravitational constant, G, to be 6.67×10^{-11} N m^2 kg^{-2}, estimate the mass of the Sun. [O*]

4. Assuming that the Earth (mass m) describes a circular orbit of radius R at angular velocity ω round the Sun (radius r, mass M) due to gravitational attraction:
(a) write down the Earth's equation of motion,
(b) Obtain the mean density of the Sun, given $\omega = 2.0 \times 10^{-7}$ rad s^{-1}; $R/r = 200$; $G = 6.7 \times 10^{-11}$ kg^{-1} m^3 s^{-2}; volume of a sphere $= \frac{4}{3}\pi r^3$. [S]

5. Explain how the mass, M, of the Sun can be calculated from a knowledge of the following:
R, distance from Earth to Sun,
r, distance from Earth to Moon,
T, orbital period of Earth,
t, orbital period of Moon,
m, the mass of the Earth. [L]

6. Show that the radius R of a satellite's circular orbit about a planet of mass M is related to its period as follows:

$$R^3 = \frac{GM}{4\pi^2} T^2$$

where G is the universal gravitational constant.

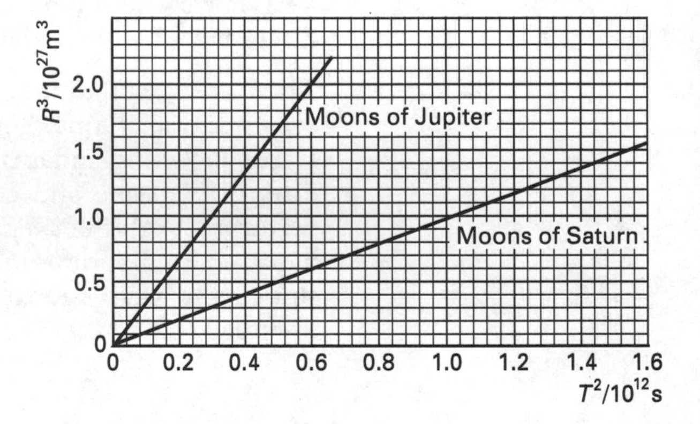

The diagram shows two graphs of R^3 against T^2; one is for the moons of Jupiter and the other is for the moons of Saturn. R is the mean distance of a moon from a planet's centre and T is its period.

The orbits are assumed to be circular.

The mass of Jupiter is 1.90×10^{27} kg.
(a) Why are the lines straight?
(b) Find a value for the mass of Saturn.
(c) Find a value for the universal gravitational constant G. [L*]

7. How did the discovery of Neptune confirm the validity of Newton's law of gravitation?

3
CLASSIFICATION OF STARS

3.1 BLACK-BODY RADIATION

A black body is a body which absorbs all the radiation that is incident on it. The concept is an idealized one, but it can be very nearly realized in practice.

Black-body radiation is the radiation emitted by a black body; its spectral distribution depends only on the temperature of the body.

Fig. 3.1 illustrates the way in which the energy radiated by a black body depends on wavelength. E_λ is such that $E_\lambda \delta\lambda$ is the energy radiated per unit time per unit surface area of the black body in the wavelength interval λ to $\lambda + \delta\lambda$. It follows that **the area under any particular curve is the total energy radiated per unit time per unit surface area at the corresponding temperature.**

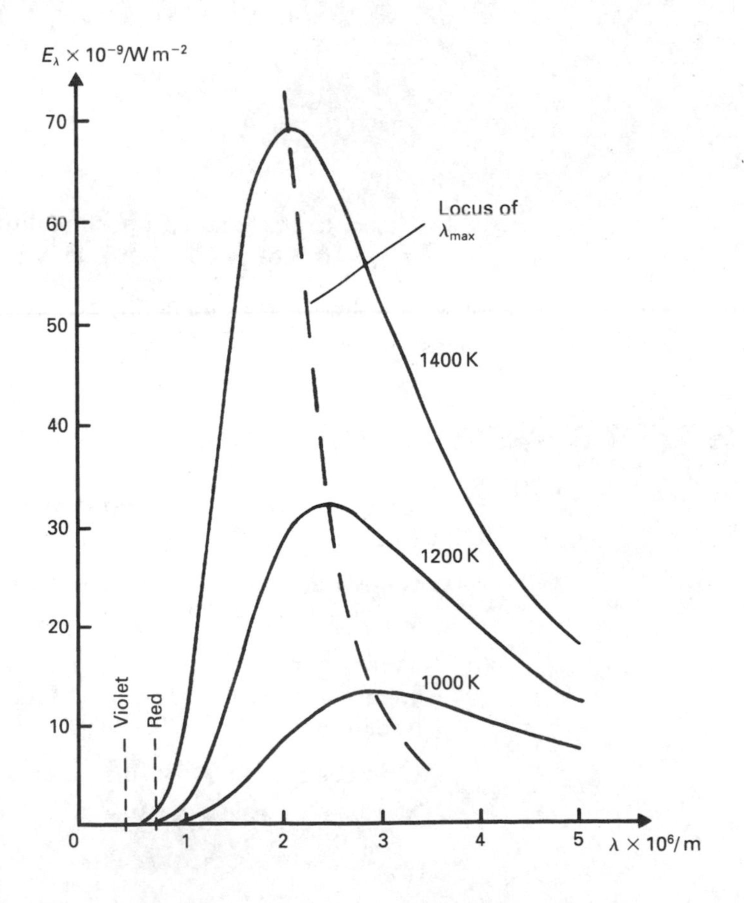

Fig. 3.1
Energy distribution of a black body

The curves illustrate the well-known observation that the colour of a body which is hot enough to be emitting visible light depends on its temperature. At about 1200 K the visible wavelengths which are emitted lie predominantly at the red end of the spectrum, and a body at this temperature is said to be red-hot. At higher temperatures the proportions of the other spectral colours increase, and the overall colour changes from red through yellow to white and finally to blue. This also accounts for the fact that different stars have different colours – the hottest stars are blue, the coolest are red and those of intermediate temperature, like the Sun, are yellow.

The curves embody two important laws.

Wien's displacement law

The wavelength λ_{max} at which the maximum amount of energy is radiated decreases with temperature and is such that

$$\lambda_{max} T = \text{a constant} \qquad [3.1]$$

Where T is the temperature of the black body in kelvins. Equation [3.1] is known as **Wien's displacement law**. The value of the constant is found by experiment to be 2.90×10^{-3} m K.

Stefan's law

The total energy radiated per unit time per unit surface area of a black body is proportional to the fourth power of the temperature of the body expressed in kelvins.

Thus

$$E = \sigma T^4$$

where

σ = a constant of proportionality known as **Stefan's constant**. Its value is $5.67 \times 10^{-8}\ \text{W m}^{-2}\ \text{K}^{-4}$.

Note that the value of E at any temperature T is equal to the area under the corresponding curve, i.e. $E = \int_0^\infty E_\lambda \, d\lambda$.

3.2 STARS AS BLACK-BODY RADIATORS

The radiation emitted by a star can be regarded as black-body radiation. It follows that:

(i) **Wien's law** can be used to estimate the surface temperature of a star once λ_{max} has been found from an examination of its spectrum,

(ii) **Stefan's law** can be used to find the radius of a star if its **luminosity** (the energy radiated per second) and its surface temperature are known, because it follows from Stefan's law that

$$\text{Luminosity} = \sigma T^4 \times \text{surface area}$$

i.e.

$$\text{Luminosity} = 4\pi r^2 \sigma T^4$$

(assuming that stars are spherical)

QUESTIONS 3A

1. The radiation emitted by the Sun has a peak intensity at a wavelength of 5.02×10^{-7} m. Calculate the temperature of the Sun's surface. (The Wien constant $= 2.90 \times 10^{-3}$ m K.)

2. The Sun has a luminosity of 3.86×10^{26} W and a surface temperature of 5.78×10^{3} K. Find the Sun's diameter.
(Stefan's constant $= 5.67 \times 10^{-8}\ \mathrm{W\,m^{-2}\,K^{-4}}$.)

3.3 APPARENT MAGNITUDE (m)

The apparent magnitude (m) of a star (or planet, galaxy, etc.) is a measure of its brightness as seen from Earth.

It is defined by

$$m = -2.5\log_{10} I + K \qquad [3.2]$$

where I is the intensity of the light from the star as measured on Earth and K is a constant.

Note The brighter a star appears to be, the lower its apparent magnitude.

If the apparent magnitudes of two stars whose intensities are I_1 and I_2 are respectively m_1 and m_2, then by equation [3.2]

$$m_1 - m_2 = (-2.5\log_{10} I_1 + K) - (-2.5\log_{10} I_2 + K)$$

$$= 2.5\log_{10} I_2 - 2.5\log_{10} I_1$$

i.e.

$$m_1 - m_2 = 2.5\log_{10}\left(\frac{I_2}{I_1}\right) \qquad [3.3]$$

It should be clear from equation [3.3] (**Pogson's formula**) that a difference of 2.5 magnitudes corresponds to an intensity ratio of 10, a difference of 5 corresponds to a ratio of 100, etc.

Notes (i) Magnitude is measured on a logarithmic scale because **the eye has a logarithmic response to light intensity**.

(ii) The Sun, the Moon and Sirius (the brightest star in the sky) have apparent magnitudes of -26.7, -12.7 and -1.47 respectively. The faintest stars that can be seen with the naked eye have magnitude 6.

3.4 THE INVERSE SQUARE LAW

A point source of light (a star is effectively a point source) emits light in all directions about the source. It follows that the intensity of the light decreases with distance from the source because the rays are spread over greater areas as the distance increases.

Rays of light at a distance d from a point source are spread over the surface of a sphere of radius d and area $4\pi d^2$. If E is the energy radiated per unit time, then the

intensity (i.e. the energy per unit time per unit area) at a distance d from the source is given by I where

$$I = \frac{E}{4\pi d^2} \qquad [3.4]$$

i.e. $$I \propto \frac{1}{d^2}$$

Thus the intensity varies as the inverse square of the distance from the source, i.e. doubling the distance reduces the intensity by a factor of 4, tripling it reduces it by a factor of 9, etc.

The inverse square law should be applied with care – it takes no account of absorption or scattering and is therefore entirely valid only in a vacuum.

QUESTIONS 3B

1. The intensity of the light that reaches the Earth from a particular star is seven times greater than that from a star whose apparent magnitude is 3.6. What is the apparent magnitude of the brighter star?

2. Find the ratio of the intensities of the light that reaches the Earth from two stars whose apparent magnitudes are **(a)** 1.4 and 3.9, **(b)** −0.72 and 4.28.

3. Find the ratio of the intensities of the light that reaches the Earth from two stars whose apparent magnitudes are 1.7 and 5.1.
(Note, if $x = \log_{10} y$, then $y = 10^x$ – use the 10^x key on your calculator.)

4. Rigel, in the constellation of Orion, has an apparent magnitude of 0.12. Find the apparent magnitude of a star which has the same intrinsic brightness as Rigel but which is four times further from the Earth.

5. The average intensity of the Sun's radiation at the surface of the Earth (after correction for absorption and scattering) is $1.37 \times 10^3\,\mathrm{W\,m^{-2}}$. Calculate **(a)** the Sun's luminosity (i.e. energy radiated per second), **(b)** its surface temperature on the assumption that it is a black body. (The mean distance of the Earth from the Sun $= 1.50 \times 10^{11}$ m, radius of the Sun $= 6.96 \times 10^8$ m, Stefan's constant $= 5.67 \times 10^{-8}\,\mathrm{W\,m^{-2}\,K^{-4}}$.)

3.5 ESTIMATING DISTANCES USING THE INVERSE SQUARE LAW

Equation [3.4] is often written in the entirely analogous form

$$F = \frac{L}{4\pi d^2}$$

where F is the **flux density** (i.e. the energy per unit time per unit area) at a distance d from the centre of a body of luminosity L. It follows that we can determine the distance to an object of known luminosity simply by measuring the flux density from it at the surface of the Earth. Examples of objects for which we can do this are Cepheid variable stars and Type 1 supernovae (see sections 4.2 and 4.4).

3.6 ABSOLUTE MAGNITUDE (M)

The absolute magnitude (M) of a star is a measure of its intrinsic brightness and is defined as the apparent magnitude the star would have if it were a distance of 10 parsecs* from the Earth.

The absolute and apparent magnitudes of a star at a distance d from the Earth are related by

$$m - M = 5\log_{10}\left(\frac{d}{10}\right) \quad [3.5]$$

where d is in parsecs.

Proof

Consider a star at a distance d (in parsecs) from the Earth. If its apparent and absolute magnitudes are m and M, then by equation [3.2]

$$m = -2.5\log_{10} I_d + K \quad [3.6]$$

and

$$M = -2.5\log_{10} I_{10} + K \quad [3.7]$$

where I_d is the actual intensity of the star, and I_{10} is the intensity it would have if it were 10 parsecs from the Earth.

Subtracting equation [3.7] from equation [3.6] gives

$$m - M = 2.5\log_{10} I_{10} - 2.5\log_{10} I_d$$

$$= 2.5\log_{10}\left(\frac{I_{10}}{I_d}\right)$$

$$= 2.5\log_{10}\left(\frac{d}{10}\right)^2 \quad \text{(by the inverse square law)}$$

$$= 5\log_{10}\left(\frac{d}{10}\right)$$

QUESTIONS 3C

1. Vega (the fifth-brightest star) has an apparent magnitude of 0.03 and is 7.7 parsecs from the Earth. Find its absolute magnitude.

2. Polaris, the Pole Star, has an apparent magnitude of 2.0 and an absolute magnitude of −4.9. Find its distance from the Earth in parsecs. (Note, if $x = \log_{10} y$, then $y = 10^x$ – use the 10^x key on your calculator.)

3. What is the difference in the absolute magnitudes of two stars if their apparent magnitudes are the same and one star is **(a)** 10 times, **(b)** 4 times farther away than the other?

4. Star A has an absolute magnitude of 2 and an apparent magnitude of 8. Star B has an absolute magnitude of 4 and an apparent magnitude of 5. **(a)** It should be immediately apparent that star A is farther away than B – how? **(b)** What is the ratio of their distances from the Earth?

*The parsec is defined in section 4.1. One parsec is approximately 3.26 light-years.

3.7 THE SPECTRAL CLASSIFICATION OF STARS

If the light emitted by a star is examined with a spectrometer, the continuous spectrum of radiation is seen to be crossed by a number of dark lines. The lines are created by the relatively cool gases in the upper layers of the photosphere (the surface layer of the star) absorbing some of the wavelengths radiated by the interior. The atoms and ions that absorb these wavelengths re-emit them, but they do so in all directions and so the radiation that reaches the spectrometer is very nearly devoid of them. **The absorption lines are characteristic of the atoms and ions that produce them** and therefore can be used to discover just which elements are present in the atmosphere of the star, and in what proportions.

Stars are classified as being of **spectral type** O, B, A, F, G, K or M on the basis of the relative intensities of the absorption lines in their spectra. Shortly after this classification had been completed, it was realized that it is actually a temperature classification, because the relative intensities of the various absorption lines in the spectrum of a star depend only on its temperature. Temperature decreases through the sequence – type O stars are the hottest and are blue, type M are the coolest and are red (Table 3.1).

Table 3.1
The main spectral types

Type	*Colour*	*Surface temperature*	*Origin of prominent absorption lines*
O	Blue	28 000 to 50 000 K	Ionized helium
B	Blue-white	10 000 to 28 000 K	Neutral helium and hydrogen
A	White	7500 to 10 000 K	Hydrogen
F	Yellow-white	6000 to 7500 K	Ca^+ and ionized metals (e.g. Fe^+)
G	Yellow	5000 to 6000 K	Ca^+ and neutral metals (e.g. Fe)
K	Orange-red	3500 to 5000 K	Molecules, Ca and neutral metals (e.g. Fe)
M	Red	2500 to 3500 K	Molecules and Ca

Note Each of the seven main types (O to M) is divided into ten divisions designated 0 to 9 with G0, for example, next to F9. The Sun is type G2, one of the hotter type G stars.

Different spectral lines are prominent at different temperatures because atomic states of ionization and excitation depend on temperature.

(i) Lines due to ionized helium occur only in the hottest (type O) stars because these are the only ones in which the temperature is high enough to cause significant numbers of helium atoms to be ionized. Since it obviously takes less energy to excite a helium atom than it does to ionize it, absorption lines due to neutral (i.e. unionized) helium are prominent in the somewhat cooler type B stars.

(ii) If hydrogen is to produce absorption lines that are in the visible part of the spectrum (the **Balmer lines**), its electron must be in the level with $n = 2$ (Fig. 3.2). Stars of types A and B produce strong hydrogen absorption lines because their temperatures are such that they have significant amounts of hydrogen in the $n = 2$ level. The hydrogen lines in cooler stars are weak because most of the hydrogen is in its ground state ($n = 1$ level). The hydrogen lines in the hottest stars are weak because most of the hydrogen is ionized. (Ionized hydrogen cannot produce absorption lines in any region of the spectrum – it has no electrons.)

Fig. 3.2 The main spectral transitions of atomic hydrogen

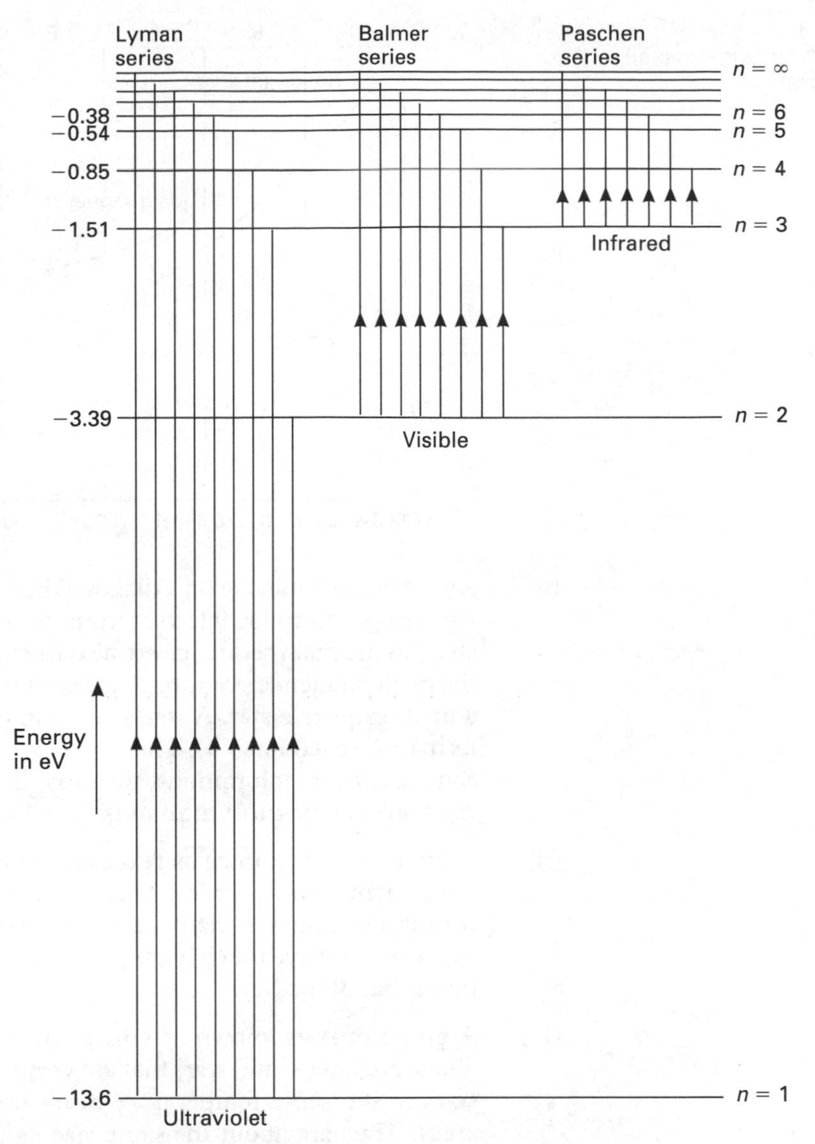

(iii) Many metals (calcium and iron for example) are more easily ionized than either hydrogen or helium, and therefore absorption lines due to ionized metals are present in the spectra of relatively cool stars. Absorption lines due to the neutral metals are observed in the spectra of even cooler stars. Metal lines are not prominent in the spectra of the hottest stars because at these temperatures the atoms are doubly, or even triply, ionized and the transitions that occur tend to be in the ultraviolet.

(iv) The coolest stars (types K and M) exhibit molecular absorption bands* because their temperatures are not so high that the molecules in them split up into their component atoms.

3.8 THE HERTZSPRUNG–RUSSELL DIAGRAM (H–R DIAGRAM)

A plot of the absolute magnitudes (or luminosities) of stars against their temperatures (or spectral types) is known as a **Hertzsprung–Russell diagram** (Fig. 3.3). The main features of it are listed overleaf.

*Atoms produce absorption lines, molecules produce bands.

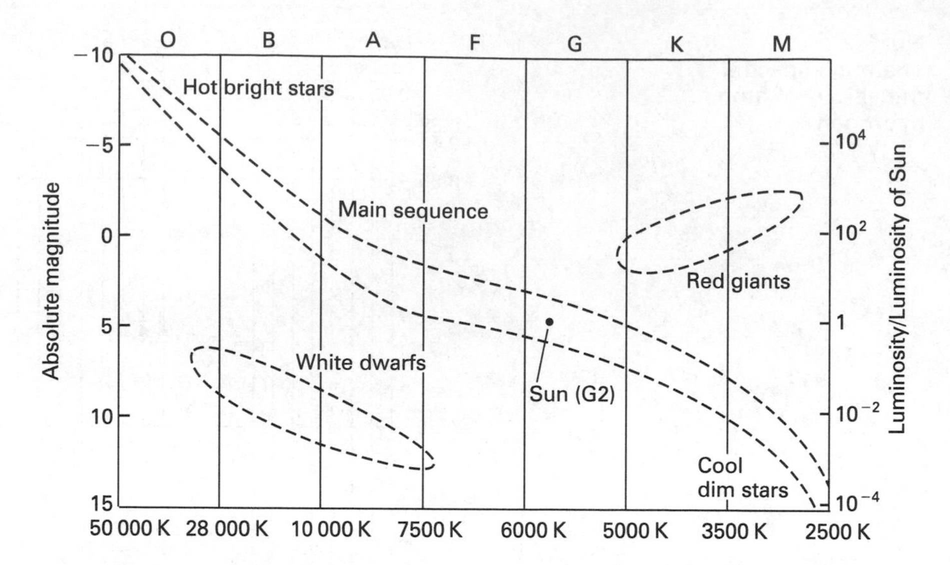

Fig. 3.3
A Hertzsprung–Russell diagram

(i) About 90% of stars lie on a diagonal band called the **main sequence**. The band slopes from top left to bottom right, reflecting the fact that the hottest stars on the main sequence are also the brightest. The diameters of most of the main-sequence stars are about the same as that of the Sun, which is itself a main-sequence star. A star's position on the main sequence depends on its mass (see section 5.9). The least massive (about 0.1 times the mass of the Sun) are the least luminous, the most massive (about 50 times the mass of the Sun) are the most luminous.

(ii) A group of stars known as **red giants** lies above the main sequence at the upper right of the diagram. These are cool (red or orange-red) stars that are very much brighter than main-sequence stars of the same temperature because they have much bigger surface areas. Their diameters are 10 to 100 times that of the Sun.

(iii) A group of stars known as **white dwarfs** lies below the main sequence. These are hot (white) stars that are very much dimmer than main-sequence stars of the same temperature because they have much smaller surface areas. They are about the same size as the Earth, i.e. their diameters are about 1/100 that of the Sun.

QUESTIONS 3D

1. Find the ratio of the luminosity of a red giant to that of the Sun if the surface temperature of the red giant is half that of the Sun and its diameter is eighty times the Sun's.

2. Find the ratio of the luminosity of a white dwarf to that of the Sun if the surface temperature of the white dwarf is three times that of the Sun and its diameter is one ninetieth the Sun's.

CONSOLIDATION

A black body is a body which absorbs all the radiation that is incident on it.

Black-body radiation is the radiation emitted by a black body; its spectral distribution depends only on the temperature of the body.

Wien's displacement law

The wavelength λ_{max} at which a black body at a kelvin temperature T radiates the maximum amount of energy is such that

$$\lambda_{max}T = \text{a constant } (= 2.90 \times 10^{-3}\,\text{m K})$$

Note λ_{max} is not the maximum wavelength emitted by the black body.

Stefan's law

The energy E radiated per unit time per unit surface area by a black body at a kelvin temperature T is given by

$$E = \sigma T^4$$

where σ is Stefan's constant ($= 5.67 \times 10^{-8}\,\text{W m}^{-2}\,\text{K}^{-4}$).

Stars can be regarded as black-body radiators and therefore for a star of radius r and surface temperature T

$$\text{Luminosity} = 4\pi r^2 \sigma T^4$$

The apparent magnitude (m) of a star (or planet, galaxy, etc.) is a measure of its brightness as seen from Earth.

The brighter a star appears to be, the lower its apparent magnitude.

If the apparent magnitudes of two stars whose intensities are I_1 and I_2 are respectively m_1 and m_2, then

$$m_1 - m_2 = 2.5\log_{10}\left(\frac{I_2}{I_1}\right) \qquad \textbf{(Pogson's formula)}$$

The absolute magnitude (M) of a star is a measure of its intrinsic brightness and is defined as the apparent magnitude the star would have if it were a distance of 10 parsecs from the Earth.

For a star at a distance d (in parsecs) from the Earth

$$m - M = 5\log_{10}\left(\frac{d}{10}\right)$$

Stars are classified as being of **spectral type** O, B, A, F, G, K or M on the basis of the relative intensities of certain absorption lines in their spectra. Temperature decreases through the sequence – type O stars are the hottest and are blue, type M are the coolest and are red.

The temperature of a star can be found by examining its spectrum to determine its spectral type or by regarding it as a black body and using Wien's law.

The absorption lines in a star's spectrum are characteristic of the atoms and ions that produce them and therefore can be used to discover just which elements are present in the atmosphere of the star, and in what proportions.

The Hertzsprung–Russell diagram (H–R diagram)

A plot of the absolute magnitudes (or luminosities) of stars against their temperatures (or spectral types). The plot reveals that there are three types of stars – main-sequence stars, red giants and white dwarfs.

QUESTIONS ON CHAPTER 3

1. **(a)** For black-body radiation, state
(i) *Wien's law,*
(ii) *Stefan's law.*
(b) The spectrum of a star has its density maximum at a wavelength of 400 nm. Assuming that the star is a perfect black body of radius 6.90×10^8 m calculate
(i) the temperature of the surface of the star,
(ii) the total radiative power output of the star.
(The Wien constant $= 2.90 \times 10^{-3}$ m K, Stefan's constant $= 5.67 \times 10^{-8}$ W m^{-2} K^{-4}.) [N*, '94]

2. State Wien's law and Stefan's law for black-body radiation.
In our Galaxy there are a number of gas clouds which emit radiation with maximum intensity at a wavelength of 10 μm. Assuming that each cloud acts as a spherical black body and that the total power output of the cloud is the same as that of the Sun, calculate for such a cloud
(a) the surface temperature,
(b) the radius.
(The Wien constant $= 2.9 \times 10^{-3}$ m K, radius of the Sun $= 7.0 \times 10^8$ m, surface temperature of the Sun $= 6000$ K.) [N*]

3. Explain what is meant by
(a) *apparent magnitude* of a star,
(b) *absolute magnitude* of a star.
State how they are related.
The table shows the absolute magnitude and apparent magnitude of the stars Rigel and Procyon.

Star	*Absolute magnitude*	*Apparent magnitude*
Rigel	−7.0	+0.1
Procyon	+2.7	+0.4

By considering the data for each star individually, deduce, without numerical calculation but giving your reasoning, which star is the furthest from the Earth. [N*]

4. Astronomers interested in a nearby star made the following measurements.
(a) The wavelength at which the intensity/wavelength graph for the star peaks.
(b) The electromagnetic energy arriving at the Earth's surface from the star per second per unit area normal to the direction of the star.
(c) The distance of the star from the Earth.
Explain how this information enables them to calculate the surface temperature of the star and its luminosity L.
How does this data, when collected together from a large number of stars, lead to the concept of a *main-sequence star*?
[L (specimen)*, '96]

5. **(a)** State in the form of equations the laws of Stefan and Wien relating to black-body radiation, defining the symbols used in the equations. Indicate what measurements would have to be made on the radiation from a star to determine **(i)** the surface temperature of the star and **(ii)** the radius of the star, given the absolute magnitude and luminosity of the Sun. What assumption has to be made in these calculation?
(b) Sketch a Hertzsprung–Russell diagram with labelled axes. Indicate on the diagram the main sequence, and the locations of giant stars and dwarf stars. By considering the radiation emitted, explain why the description 'dwarf' is appropriate. [N]

6. **(a)** **(i)** Explain what is meant by a *black body*. Using the same axes draw sketch graphs of the intensity of the radiation emitted by a black body as a function of wavelength for **three** widely different temperatures. Label your graphs with the temperatures and indicate an approximate scale on the wavelength axis.
(ii) Stars may be arranged according to their colours into different spectral classes. Use the graph you have drawn in **(i)** to draw conclusions about the surface temperatures of stars in the different spectral classes.
(b) Although all visible stars are roughly uniform in composition and all consist mostly of hydrogen and helium, the absorption spectra of stars differ from each other, depending on the spectral class.
(i) Explain how absorption spectra are produced in stellar spectra.
(ii) Explain why only some spectral classes show dark lines characteristic of metallic elements in their absorption spectra. [N*]

7. **(a)** **(i)** Sketch a Hertzsprung–Russell diagram for stars, with absolute photographic magnitude as the vertical axis and temperature as the horizontal axis. Indicate values on each axis. On the diagram mark the positions of the main-sequence stars, the Sun, red giant stars and white dwarf stars. In addition, mark on the horizontal axis the spectral classes O to M of the stars.

(ii) Use your diagram to explain why red giant stars must be much larger than white dwarf stars.

(b) Describe the principal spectral features of each of the spectral classes O to M.

[N, '93]

8. **(a)** Explain what is meant by the *apparent magnitude* of a star.

(b) Given that a first magnitude star is 100 times brighter than a sixth magnitude star, obtain an expression relating the ratio of the apparent brightness of two stars to the difference in their apparent magnitudes.

(c) The variable star R R Lyrae has an apparent magnitude ranging from 7.1 to 7.8. Use the expression you have obtained in **(b)** to determine the ratio of the maximum apparent brightness to the minimum apparent brightness for this variable star. [N*]

9. **(a)** Explain what is meant by the *luminosity* of a star.

'The ratio of the luminosity of a first magnitude star to that of a sixth magnitude star is 100 to 1.'

From this statement calculate the ratio of the luminosity of a first magnitude star to that of a second magnitude star.

(b) The Sun has an apparent magnitude of −25 when seen from a distance of two astronomical units. Use the definition of the absolute magnitude of a star, in terms of its apparent magnitude and distance, to calculate the absolute magnitude of the Sun.

(1 parsec = 206 300 astronomical units.)

[N]

10. **(a)** The surface temperature T of the Sun is 5800 K and its radius is 6.96×10^8 m. Calculate a value for its luminosity L given that

$$L = \sigma T^4 \times \text{surface area}$$

where $\sigma = 5.67 \times 10^{-8}\,\mathrm{W\,m^{-2}\,K^{-4}}$.

How, in principle, may the temperature of a distant main-sequence star be determined?

How can the radius of a distant main-sequence star be estimated once its temperature is known?

(b) Show on a simple sketch of a Hertzsprung–Russell diagram, and in approximately the correct positions, a red giant star labelled R, a high-mass main-sequence star labelled M and a white dwarf star labelled W.

With reference to their coordinates on your diagram, explain the physical significance of the descriptive titles *red giant* and *white dwarf*. [L*, '96]

4

DETERMINING STELLAR DISTANCES

4.1 TRIGONOMETRIC PARALLAX

Trigonometric parallax is a method of determining stellar distances by using the surveying technique of triangulation. It is based on the fact that nearby stars appear to move relative to more distant ones as the Earth orbits the Sun.

The annual parallax (p) of a star is half the angle through which the direction of the star shifts as the Earth moves from one side of its orbit to the other (Fig. 4.1).

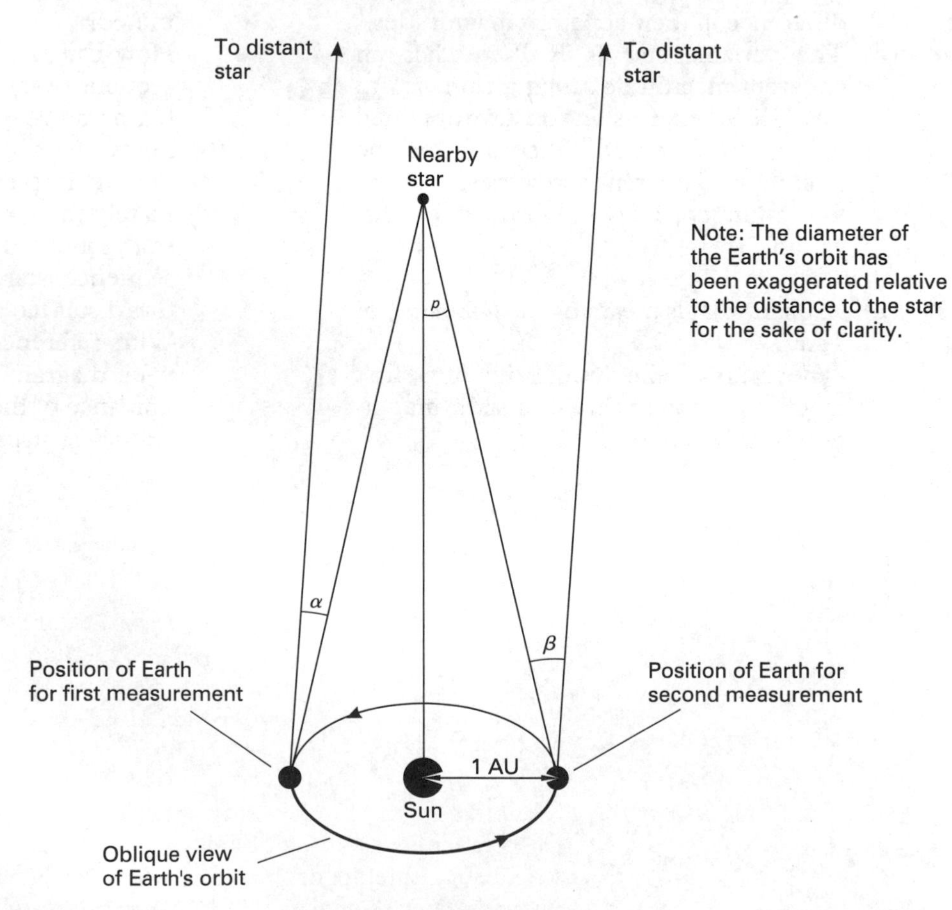

Fig. 4.1 Diagram to illustrate annual parallax

The parsec (pc) is the distance from the Earth to a point that has an annual parallax of 1 arc second (i.e. 1/3600 of a degree) – see Fig. 4.2.

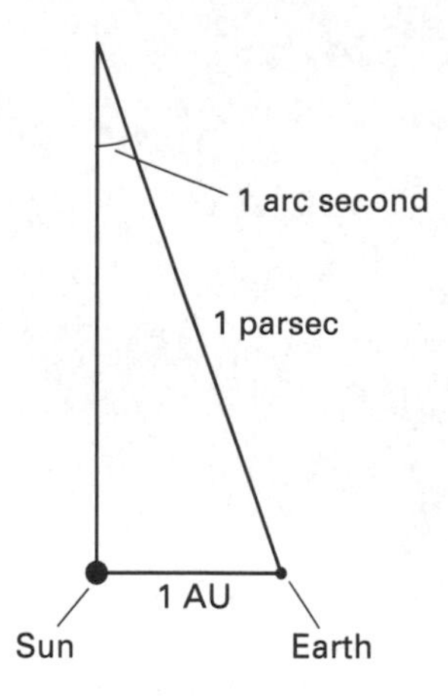

Fig. 4.2
Diagram to illustrate the parsec

It should be clear from Fig. 4.1 that **the closer a star is to the Earth, the bigger its annual parallax**. It can be shown that, in general

$$\text{Distance from Earth in parsecs} = \frac{1}{\text{Annual parallax in arc seconds}} \quad [4.1]$$

It is the simplicity of this relationship that is the rationale behind the definition of the parsec. Its validity rests on the fact that the small angle approximation ($\sin\theta \approx \tan\theta \approx \theta$) applies because the largest annual parallax is a mere 0.76 arc seconds (for Proxima Centauri, the nearest star to the Earth).

The annual parallax of a star is measured by determining the angle between the star and a much more distant one (i.e. one that does not appear to move) on two occasions six months apart – if the first angle is α and the second is β (Fig. 4.1), then

$$p = \frac{\alpha + \beta}{2}$$

In practice, α and β are found by measuring the displacements of the star from a more distant star on photographs taken six months apart.

A correction has to be applied to take account of the **proper motion** of the star, i.e. the number of arc seconds per year by which its position changes relative to the more distant stars because of its actual motion through space. This can be distinguished from the (apparent) motion due to parallax simply because it is always in the same direction – that due to parallax cycles back and forth over the course of a year.

The errors involved in measuring annual parallax are such that values of less than about 0.01 arc seconds are not reliable. The method can be used, therefore, only for stars which are closer than about 100 parsecs. This is not very far in astronomical terms – only about 1% of the distance to the centre of our own Galaxy!

Notes (i) Refraction by the Earth's atmosphere limits the accuracy of Earth-based measurements. There is obviously no such problem with measurements made from a satellite orbiting above the atmosphere. Data collected by the European Space Agency satellite, **Hipparcos**, which operated from 1989 to 1993, should yield parallax values extending out to about 1000 parsecs when it is finally processed.

Fig. 4.3
Diagram for Note (ii)

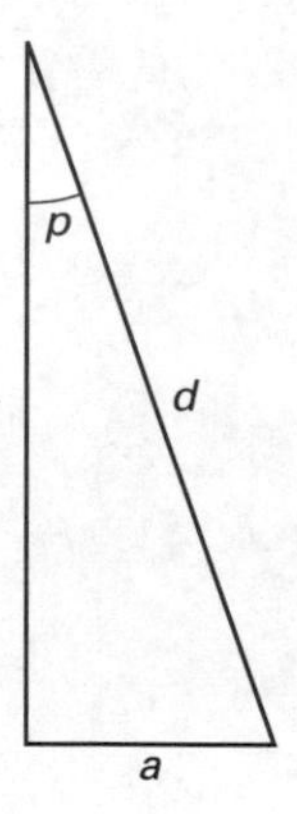

(ii) Refer to Fig. 4.3. Since p is a small angle

$$p \text{ (in radians)} = \frac{a}{d}$$

Therefore, since

$$1 \text{ radian} = \frac{360}{2\pi} \times 3600 = 2.063 \times 10^5 \text{ arc seconds}$$

$$p \text{ (in arc seconds)} = 2.063 \times 10^5 \times \frac{a}{d}$$

When $p = 1$ arc second and $a = 1$ AU, $d = 1$ parsec, and therefore

$$1 = 2.063 \times 10^5 \times \frac{1 \text{ AU}}{1 \text{ parsec}}$$

i.e.

$$1 \text{ parsec} = 2.063 \times 10^5 \text{ AU} = 3.262 \text{ light-years}$$

QUESTIONS 4A

1. The first person to succeed in measuring the distance to a star by parallax was Friedrich Bessel who measured the parallax of 61 Cygnus in 1838. How far from the Earth is 61 Cygnus **(a)** in parsecs, **(b)** in light-years if its annual parallax is 0.29 arc seconds?

4.2 CEPHEID VARIABLES

The Cepheid variables belong to a particular class of variable stars whose luminosity varies (typically by about one magnitude) in an extremely regular manner, with periods ranging from 1 to 100 days. **All Cepheid variables brighten more quickly than they fade**. The light curve of δ (Delta) Cephei (Fig. 4.4), the first of the Cepheids to be discovered, is typical.

There is a simple relationship between the absolute magnitudes of the Cepheids and their periods – the period–luminosity law (Fig. 4.5). We can therefore find the absolute magnitude (M) of any Cepheid simply by measuring its

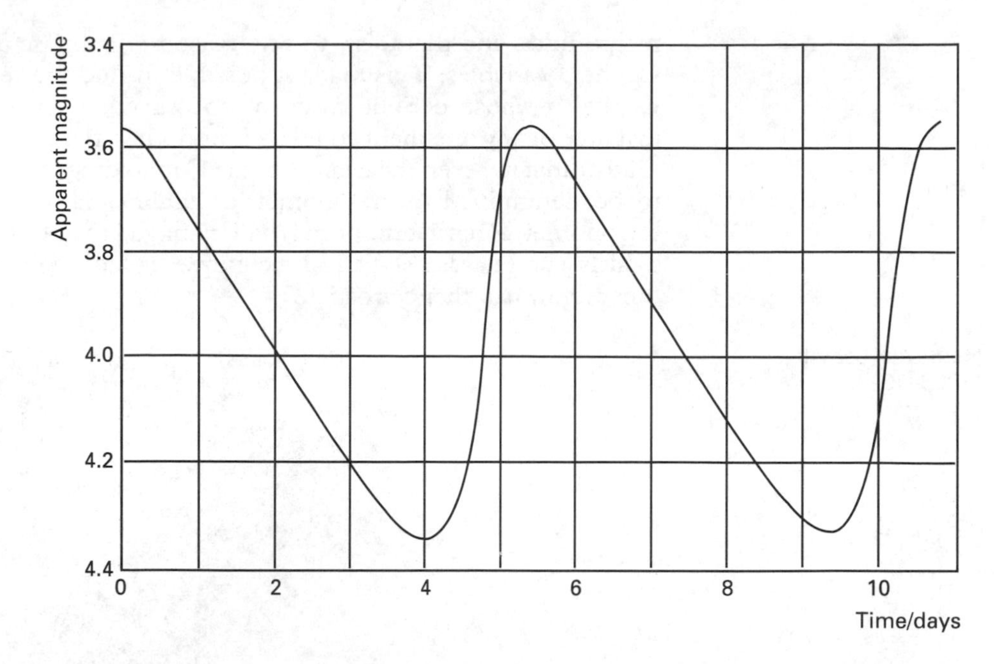

Fig. 4.4
The light curve of δ (Delta) Cephei

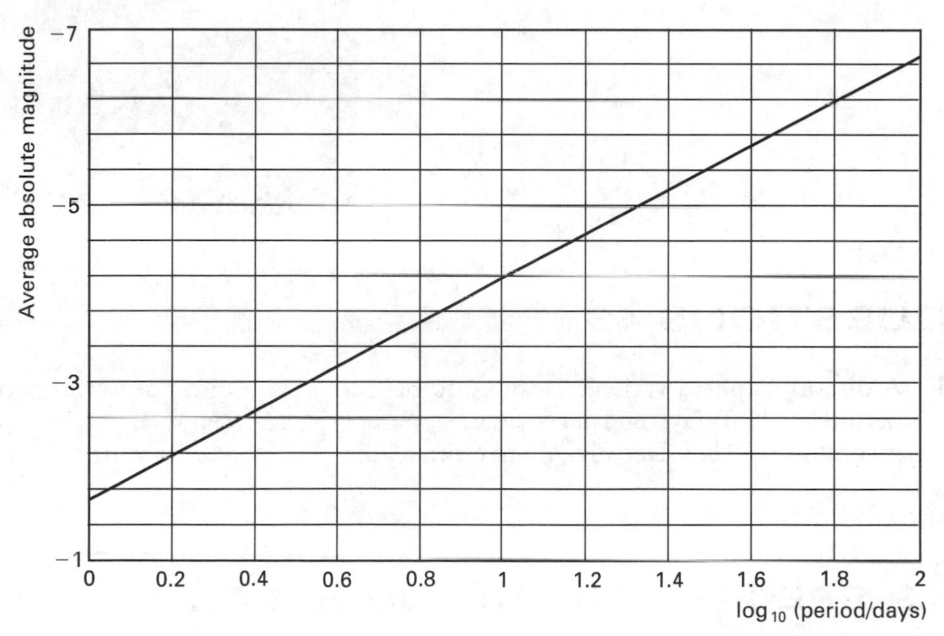

Fig. 4.5
The period–luminosity law

period. If we then measure its apparent magnitude (m), we can find its distance (d) from

$$m - M = 5\log_{10}\left(\frac{d}{10}\right) \qquad \text{(equation [3.5])}$$

Cepheids are very bright stars (F–K supergiants, typically 10 000 times brighter than the Sun). They can therefore be seen at great distances and allow us to determine the distances to galaxies as much as 5×10^6 parsecs away.

The period–luminosity law was discovered in 1912 by Henrietta Leavitt, who noticed that there was a simple relationship between the periods and the apparent magnitudes of 25 Cepheids in the Small Magellanic Cloud. Since, to a good approximation, all 25 stars are at the same distance from the Earth, she realized that there must be a similar relationship between the periods and the absolute magnitudes, and that there was no reason to suppose that it did not apply to all Cepheid variables. This made it possible to find the relative distances of all the Cepheids whose periods could be measured. It also meant that if the actual distance of any one them could be found, then the distances of all of them could. Unfortunately, even the nearest of the Cepheids is far too far away for its distance to be determined by trigonometric parallax. However, by 1913, the Danish astronomer, Ejnar Hertzsprung, had managed to find the distance of one by using a fairly complicated statistical technique called main-sequence fitting – astronomers now had their yardstick!

Henrietta Leavitt
(1868–1921)

QUESTIONS 4B

1. A distant Cepheid variable is found to have a period of 25.0 days and an average apparent magnitude of 18.5. Use Fig. 4.5 to estimate the distance to this particular Cepheid.
(Note, if $x = \log_{10} y$, then $y = 10^x$ – use the 10^x key on your calculator.)

4.3 SPECTROSCOPIC PARALLAX

The spectral type (e.g. A2, K7, etc.) of any star can be found from an analysis of its spectrum (section 3.7). The spectrum also allows us to determine whether any particular star is a main-sequence star, a white dwarf or a red giant etc. The absorption lines in a white dwarf, for example, are much broader than those in a main-sequence star of the same spectral type. (This is a consequence of the atoms in a white dwarf being much closer together than those of a main-sequence star.) Armed with this information, we can plot the star's approximate position on an H–R diagram (section 3.8) and then simply trace across to find its absolute magnitude (M). Providing we also know its apparent magnitude (m), we can find its distance (d) from

$$m - M = 5\log_{10}\left(\frac{d}{10}\right) \qquad \text{(equation [3.5])}$$

Distances obtained by spectroscopic parallax are not very precise – absolute magnitude values cannot be determined to better than about ±1 because of the large amount of vertical scatter on the H–R diagram. Despite this, it is the most common method of estimating stellar distances, because it is often the only means available. It can be used for any star that is bright enough for us to be able to obtain its spectrum. The upper limit is of the order of 10^5 parsecs.

4.4 SUPERNOVAE

All Type 1 supernovae have an absolute magnitude of −20 at the peak of their outburst. It follows that the distance of a galaxy in which a supernova occurs can be found by measuring the maximum apparent magnitude of the supernova and using equation [3.5]. The peak output is equivalent to that of an entire galaxy and can be detected at distances of over 10^8 parsecs.

CONSOLIDATION

The annual parallax (*p*) of a star is half the angle through which the direction of the star shifts as the Earth moves from one side of its orbit to the other.

The parsec (pc) is the distance from the Earth to a point that has an annual parallax of 1 arc second (i.e. 1/3600 of a degree).

$$\text{Distance from Earth in parsecs} = \frac{1}{\text{Annual parallax in arc seconds}}$$

1 parsec $= 2.063 \times 10^5$ AU $=$ 3.262 light-years

Trigonometric parallax can be used to find the distances of stars up to 100 parsecs from Earth.

Cepheid variables can be used to find the distances of stars up to 5×10^6 parsecs from Earth.

All Cepheid variables brighten more quickly than they fade.

There is a simple relationship between the absolute magnitudes and the periods of the Cepheid variables – the **period–luminosity law**.

QUESTIONS ON CHAPTER 4

1. Explain the principles underlying the trigonometrical parallax method of stellar-distance measurement, and show how the distance of the star is obtained. State **two** corrections which have to be made to the calculated distance. [N*, '91]

2. (a) Define the *parsec* and explain the reason for the annual parallax of a star.
   A sequence of photographs of the same group of stars is taken, with the aid of a telescope and camera attachment, at regular intervals during a period greater than one year. On the photographs, all the star images can be superposed except for one. This image is displaced on consecutive photographs up to a maximum, but returns to its original position when the time interval between photographs is one year. Describe how the information on the photographs can be used to determine the distance of this star, explaining carefully each stage in the process.
   (b) Explain what is meant by a *Cepheid variable* star, and show how measurements on these stars enable the distances of the nearer galaxies to be determined. [N]

3. **(a)** Explain what is meant by
(i) *apparent magnitude* of a star,
(ii) *absolute magnitude* of a star.
(b) State what you understand by *spectral class* of a star. Outline the method of spectroscopic parallax for determining the distance of a star. Explain what astronomical data must be known before this method can be used.
(c) **(i)** Sketch a graph showing the variation of apparent magnitude as a function of time for Cepheid variables.
(ii)

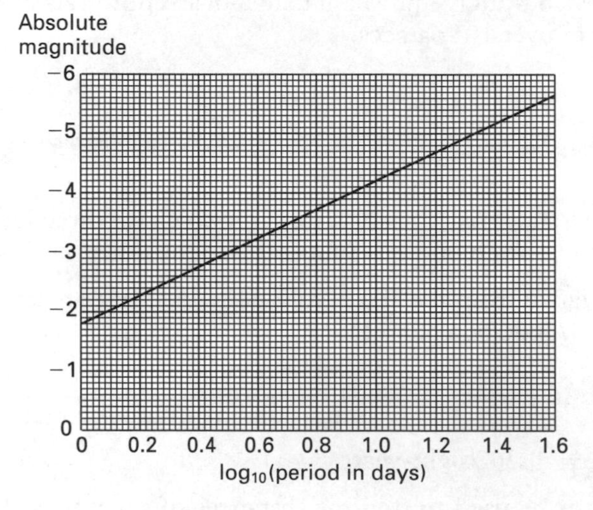

The figure shows the relationship between the absolute magnitude and period of pulsation for Cepheid variables. Given that Delta Cephei has a period of pulsation of 5.2 days and that the mean value of its apparent magnitude if 3.9, use the data in the figure to estimate the distance, in parsec, of Delta Cephei from Earth.
[N, '90]

5

THE LIVES OF STARS

5.1 STAR FORMATION

Stars form from large clouds of interstellar gas and dust. (The gas is mainly hydrogen, with some helium and trace amounts of heavier elements.) Some regions of these clouds are more dense than others, and one that is dense enough, and cool enough, will contract under the effect of its own gravity. As it does, it draws in more and more material from the rest of the cloud to form what is known as a **protostar**. The contraction continues, and the protostar heats up as gravitational potential energy is converted to kinetic energy. The temperature in the interior eventually becomes high enough ($\sim 10^7$ K) for thermonuclear fusion reactions to start, releasing enormous amounts of energy. The contraction ceases as soon as a **hydrostatic equilibrium** is established in which the sum of the thermal and radiation pressure from the interior is equal to the gravitational pressure. Finally, a **temperature equilibrium** is established in which the rate at which energy is produced in the core is balanced by the rate at which it is radiated away. The protostar is now a stable star, about to begin the main-sequence phase of its existence.

Notes

(i) **Stars spend only a tiny fraction of their lives as protostars**. The Sun, for example, will spend 10 billion years as a main-sequence star but took only about one million years to form.

(ii) Protostars whose masses are less than about 0.08 times the mass of the Sun can never become hot enough for nuclear fusion to begin and so become low-temperature pseudo-stars known as (the still to be discovered) **brown dwarfs**.

5.2 THE FUSION PROCESS

The energy radiated by a star is generated by thermonuclear fusion reactions in its core. The process that occurs in main-sequence stars is known as **hydrogen burning** – the thermonuclear conversion of hydrogen to helium. In a star whose mass is less than or equal to the Sun's, it proceeds by way of a process called the **proton–proton chain**; in heavier stars it occurs primarily through a series of reactions known as the **CNO cycle**. The net result is the same in each case – four protons combine to produce a helium nucleus together with two positrons (e^+) and two neutrinos (ν).

$$4\,^1H \rightarrow {}^4He + 2e^+ + 2\nu$$

The proton–proton chain

This is a three-stage process.

1 Two protons fuse to produce a deuterium nucleus:

$$^1H + {}^1H \rightarrow {}^2D + \nu + e^+$$

2 A third proton fuses with the deuterium to produce a helium 3 nucleus:

$$^1H + {}^2D \rightarrow {}^3He$$

3 Two helium 3 nuclei fuse to produce a helium 4 nucleus and two protons:

$$^3He + {}^3He \rightarrow {}^4He + {}^1H + {}^1H$$

Stage 3 requires stages 1 and 2 to occur twice and therefore a total of six protons are used. Since stage 3 releases two protons, the net result is that four have been used, i.e.

$$4 \text{ protons} \rightarrow 1 \text{ helium 4 nucleus} + 2 \text{ positrons} + 2 \text{ neutrinos}$$

It can be shown that the reaction releases 6.3×10^{14} joules of energy for every kilogram of protons consumed. It follows from $E = mc^2$ that the associated decrease in mass is 0.007 kg. It is estimated that the mass of hydrogen in a newly formed star that is available for fusion (i.e. that in the core of the star, where the temperature is high enough for fusion to occur) amounts to about 10% of the total mass of the star. We are now in a position to estimate the lifetime of the Sun.

$$\text{Mass of Sun} = 2.0 \times 10^{30}\text{ kg}$$

$$\therefore \quad \text{Mass of hydrogen available for fusion} = 2.0 \times 10^{29}\text{ kg}$$

$$\therefore \quad \text{Total energy output} = 2.0 \times 10^{29} \times 6.3 \times 10^{14} = 12.6 \times 10^{43}\text{ J}$$

The Sun radiates energy at a rate of $3.9 \times 10^{26}\text{ J s}^{-1}$, and therefore

$$\text{Lifetime} = \frac{12.6 \times 10^{43}}{3.9 \times 10^{26}} = 3.2 \times 10^{17}\text{ s} = 1.0 \times 10^{10}\text{ years}$$

This is in good agreement with the age of the Universe, thought to be about 1.5×10^{10} years. Geological evidence suggests that the Earth is 4.6×10^9 years old. According to current theories, the Sun came into being at the same time as the rest of the solar system and therefore appears to be about halfway through its life.

If two protons are to fuse, they must get close enough for the short-range, attractive force (called the **strong interaction** or **nuclear force**) to overcome their mutual electrostatic repulsion. According to classical physics the protons need to have sufficient kinetic energy to do this. The kinetic theory gives the average KE of a particle of a gas at a kelvin temperature, T, as $\frac{3}{2}kT$, where k is Boltzmann's constant ($= 1.38 \times 10^{-23}\text{ J K}^{-1}$). The electrostatic PE of two protons whose separation is r is $e^2/(4\pi\varepsilon_0 r)$ where e is the charge on the proton and ε_0 is the permittivity of vacuum. Thus, according to classical physics we should be able to estimate the temperature required for fusion by putting

$$\tfrac{3}{2}kT = \frac{e^2}{4\pi\varepsilon_0 r} \qquad [5.1]$$

with $r = 4 \times 10^{-15}$ m (the distance at which the strong interaction might be expected to take over from electrostatic repulsion). If we do this, we find $T \approx 3 \times 10^9$ K – which is about 200 times the temperature (1.5×10^7 K) at the core of the Sun! There are two reasons why fusion occurs at a much lower temperature than that predicted by equation [5.1].

(i) The kinetic theory gives the average kinetic energy of the protons – a small proportion of the protons will have the required energy.

(ii) The protons can **tunnel** through the electrostatic potential barrier; there is no need for them to climb over it. According to quantum mechanics a proton can be thought of as a wave. The probability of finding the proton at any particular point is proportional to the square of the amplitude of the wave at that point. The amplitude is very small in the barrier region (i.e. where the energy of the proton is less than the electrostatic PE) but it is not zero, and there is therefore a small but finite probability of the proton passing through it.

A simple calculation reveals that over the course of any one second only about 1 in 10^{17} of the protons in the core of the Sun is taking part in a fusion reaction. This is because, even with tunnelling, the probability of any particular proton achieving the first stage of the proton–proton chain is extremely low – it takes about 5 billion years on average. It is fortunate that the process is so slow – this is what allows the Sun to release its energy steadily rather than in one violent explosion. Once a deuteron (a deuterium nucleus) has been created, the second stage occurs in under a second. The average time that a helium 3 nucleus has to wait before accomplishing the third stage of the chain is about 1 million years.

The CNO cycle

This occurs in stars that are heavier than the Sun and where the core temperature exceeds 1.6×10^7 K. It is a six-stage process in which a carbon 12 nucleus acts as a catalyst:

1 $^{1}H + ^{12}C \rightarrow ^{13}N$

2 $^{13}N \rightarrow ^{13}C + e^{+} + \nu$ (positron decay)

3 $^{1}H + ^{13}C \rightarrow ^{14}N$

4 $^{1}H + ^{14}N \rightarrow ^{15}O$

5 $^{15}O \rightarrow ^{15}N + e^{+} + \nu$ (positron decay)

6 $^{1}H + ^{15}N \rightarrow ^{12}C + ^{4}He$

The net result is that 4 protons have combined to form a helium 4 nucleus plus two positrons and two neutrinos.

5.3 ENERGY TRANSFER IN STARS

The energy produced in the core of the Sun (or any other main-sequence star) reaches the surface by radiative diffusion and by convection.

Radiative diffusion occurs in a 300 000 km thick region surrounding the Sun's core known as the **radiative zone** (Fig. 5.1). Photons created in the core slowly work their way through the radiative zone by a random-walk process. They are absorbed by atoms and are then re-emitted, only to be absorbed again after travelling an average of about 1 centimetre. They are, of course, re-emitted in all directions, but with a slight tendency to move away from the core rather than towards it because of the temperature gradient, i.e. the net flow is from hotter regions to cooler regions. The fact that the photons travel only a short distance between successive absorptions makes the process an extremely slow one – it takes about 1 million years for the energy to make its way from the core to the surface.

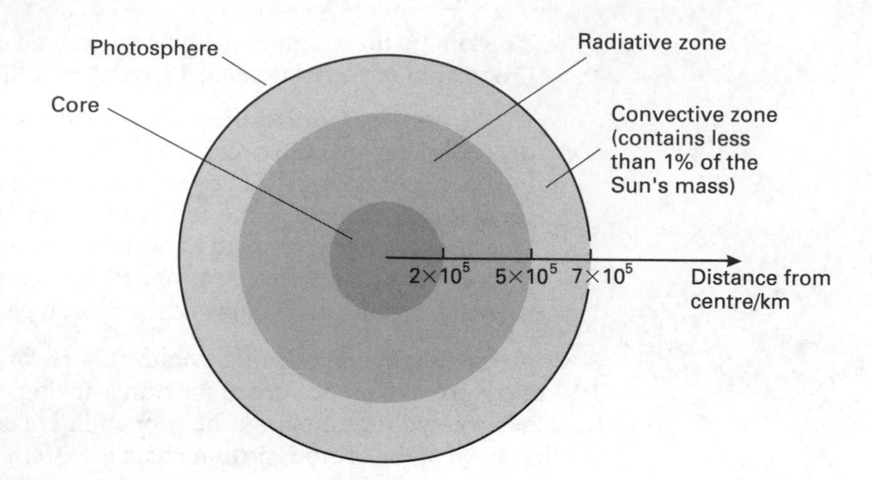

Fig. 5.1
Internal structure of the Sun

On average, each re-emission occurs in a region that is at a slightly lower temperature than that where the previous one occurred. This has the effect of degrading the photon energies – what starts out as a high-energy γ-ray photon ends up as thousands of optical photons.

The convective zone is a 200 000 km thick layer surrounding the radiative zone where conditions are such that **convection** is the most effective means of energy transfer. Energy is carried outwards as hot gases rise towards the surface and are replaced by cooler gases moving down from the surface.

5.4 EVOLUTION AFTER THE MAIN SEQUENCE

A star stays on the main sequence until the hydrogen in its core is exhausted, by which time the core is very nearly pure helium. There is no longer any radiation pressure to maintain the size of the core, and it shrinks under gravity, heating up in the process and allowing hydrogen burning to take place in a spherical shell surrounding the shrunken core. The outer layers expand far into space but the core continues to contract and its temperature continues to rise. As the outer layers expand, the surface cools to about 3500 K – the star is now a **red giant**, and its diameter is 10 to 100 times that of the original star.

When the temperature in the core reaches 10^8 K, **helium burning** begins – a process in which helium nuclei fuse to produce carbon or oxygen. (It requires a higher temperature than hydrogen burning because helium nuclei have twice as many protons and therefore four times the electrostatic repulsion.)

Initially, two helium nuclei fuse to produce beryllium:

$$^{4}\text{He} + {}^{4}\text{He} \rightarrow {}^{8}\text{Be}$$

Beryllium 8 is highly unstable and breaks back down into two helium nuclei unless it manages to fuse with a third helium nucleus before it has a chance to. This second reaction produces a stable isotope of carbon:

$$^{4}\text{He} + {}^{8}\text{Be} \rightarrow {}^{12}\text{C}$$

The process is known as the **triple–alpha process**. This almost simultaneous fusion of three helium nuclei is a low-probability event, and therefore once sufficient carbon has been formed the helium is more likely to be consumed by fusing with carbon to produce a stable isotope of oxygen:

$$^{4}\text{He} + {}^{12}\text{C} \rightarrow {}^{16}\text{O}$$

A red giant burns helium for about 20% of the time it spent burning hydrogen as a main-sequence star. When all the helium is used up the core starts to contract under gravity, and its temperature starts to rise again. The remainder of the star's evolution depends on its mass.

The core of a red giant whose mass is less than about 4 solar masses never reaches a temperature high enough to initiate any further fusion reactions. Instead, the star becomes unstable and sheds its outer layers, exposing the carbon–oxygen core. Radiation from the core excites the ejected gas and causes it to glow as a so-called **planetary nebula**. The core itself appears as a (burned out) extremely hot blue star, which eventually cools and contracts to form a **white dwarf**.

The planetary nebula M57 in the constellation of Lyra. The star that ejected this gas and which is causing it to glow can be seen as a white dot at its centre.

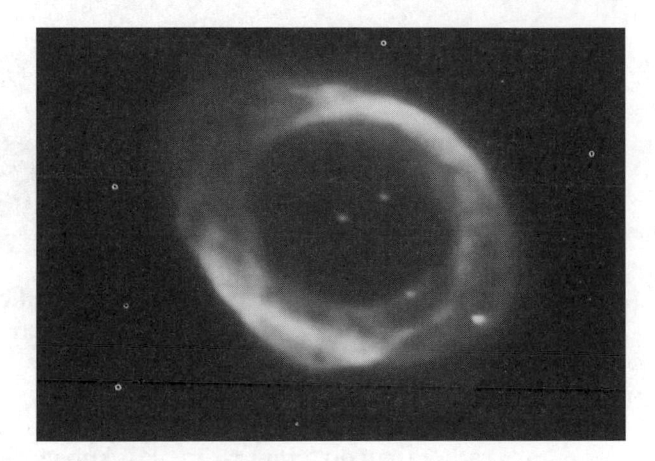

The temperature in the core of a star whose mass is at least 4 times that of the Sun becomes high enough to start fusion reactions that produce elements heavier than carbon. The more massive the star, the higher its core temperature can become and the heavier the nuclei it is able to manufacture. The heaviest is iron, in the form of the most stable of all isotopes, ^{56}Fe. Fusion cannot continue beyond iron because it would not release energy. Once fusion has ceased, the core collapses catastrophically and the outer layers of the star are thrown off in a (Type II) **supernova** explosion. The core is left behind as a **neutron star** or as a **black hole**.

Note Elements that are heavier than iron can be produced in neutron capture reactions in stars that have a plentiful supply of neutrons and in supernova explosions. (These reactions are endothermic, i.e. they absorb energy rather than release it, and therefore cannot act as the energy source of the star or supernova.) A nucleus does not become the nucleus of a different element simply by absorbing a neutron, merely a different isotope of the same element. However, if the new nucleus is unstable and decays by (negative) β-emission, its atomic number (proton number) increases by 1, i.e. it becomes the nucleus of a heavier element. Successive reactions of this type eventually build up the entire periodic table of elements. It is generally believed that the heaviest elements are produced exclusively in supernovae.

5.5 WHITE DWARFS

As we explained in section 5.4, a white dwarf is the final stage in the evolution of a low-mass star. They are about the size of the Earth but have the mass of the Sun, and consequently have enormous densities – typically $10^9\ \mathrm{kg\,m^{-3}}$. The first white dwarf to be recognized as such was Sirius B; it has the same mass as the Sun but has 100 000 times its density and a surface temperature of 30 000 K. Because of their small size, they are very dim and none can be seen with the naked eye.

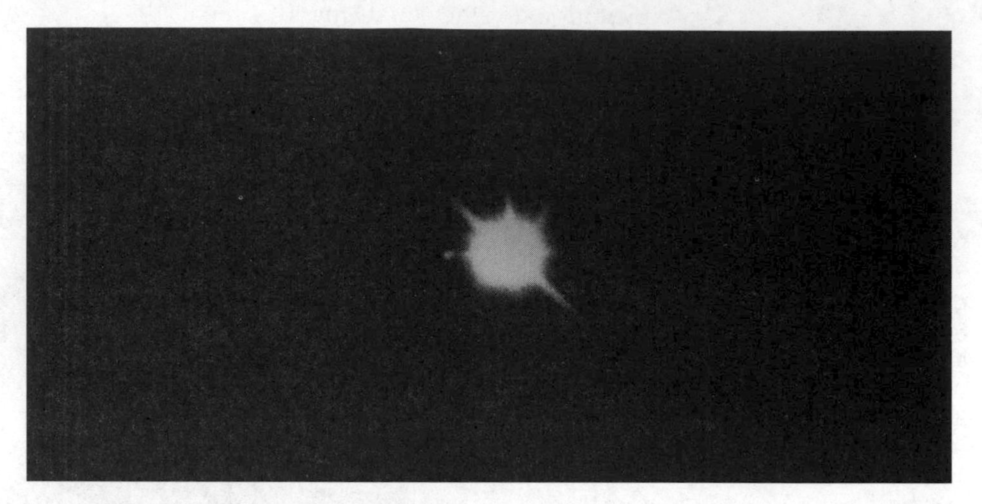

Sirius (the brightest star in the sky) and its white-dwarf companion, Sirius B, seen here at the 9 o'clock position

A white dwarf consists mainly of ionized carbon and oxygen atoms packed tightly together in a sea of electrons. The electrons are so closely packed that quantum effects become important. Electrons are subject to the **Pauli exclusion principle**, and therefore two electrons cannot be in the same quantum state. The effect of this is to make the electrons exert a pressure that depends on density rather than on temperature. It is known as **electron degeneracy pressure** and it is this that prevents the white dwarf collapsing under its own gravity.

It is impossible for a white dwarf to have a mass of more than 1.4 solar masses (the **Chandrasekhar limit**). If the mass exceeds this value, electron degeneracy pressure is incapable of resisting the gravitational pressure and the star collapses to form either a neutron star or a black hole.

5.6 NEUTRON STARS

A neutron star is a small, extremely dense star composed almost entirely of neutrons. (A neutron star of 1.4 solar masses has a diameter of only 10 km and a density of the order of 10^{15} kg m^{-3} – a million times that of a white dwarf.) They are formed when the core of a massive, dying star becomes so compressed that electrons are forced inside nuclei and combine with protons to produce neutrons. The nuclei cease to exist independently and merge together creating a sea of free nucleons, most of which are neutrons. They exert a pressure, akin to electron degeneracy pressure, known as **neutron degeneracy pressure**.

There is an upper limit to the mass of a neutron star because there comes a point, as the mass increases, where neutron degeneracy pressure is unable to resist the increasing gravitational pressure. This is equivalent to the Chandrasekhar limit of a white dwarf, and is thought to be between 2 and 3 solar masses.

Pulsars

Neutron stars have intense magnetic fields (typically 10^{12} times that of the Earth) and spin very rapidly (up to 30 rev s^{-1}). This incredible rate of rotation is a consequence of the conservation of angular momentum – as the original (slowly spinning) star collapses to form a neutron star, it spins faster in order to conserve angular momentum, just as a spinning skater speeds up when she pulls her arms in towards her body. The rotating magnetic field creates an electric field that accelerates electrons along the magnetic axis, causing two narrow beams of electromagnetic radiation, usually in the form of radio waves, to be emitted from the star's north and south magnetic poles (Fig. 5.2). Because the magnetic axis is at an angle to the axis of rotation, each beam sweeps out the surface of a cone as the star rotates. If the Earth is in the path of one of the beams, we detect it as a short pulse of radiation – in much the same way that the beam from a lighthouse appears to flash on and off.

Fig. 5.2
A pulsar

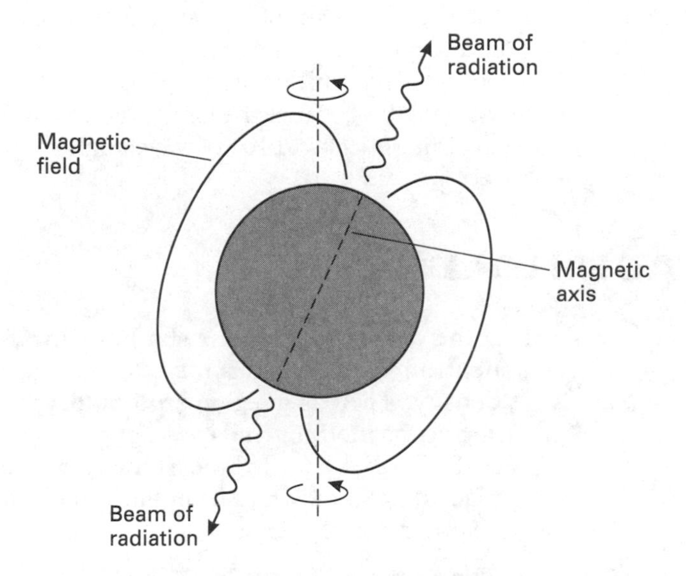

The first known pulsar was discovered by Jocelyn Bell in 1967 when she detected radio pulses with a period of 1.337 301 1 seconds that was constant to better than one part in a million. About 500 have been discovered since, with periods ranging from 0.03 to 4 seconds.

Jocelyn Bell photographed in 1977

In 1968, a pulsar was discovered at the centre of the **Crab Nebula** – the expanding remnant of a supernova seen by Chinese astronomers in 1054. This particular discovery, which clearly connected pulsars with supernovae, was regarded as strong confirmation of the theory that neutrons stars can be formed in supernova explosions. The majority of supernova remnants (SNRs) have no observable pulsar. This should not surprise us – we cannot be aware of the presence of a pulsar unless the Earth happens to intercept one or other of its rotating beams, and not all supernova explosions produce a neutron star in the first place. Nor should it surprise us that when we do detect a pulsar, it is only rarely that we also find an associated SNR – most pulsars are over a million years old and the SNR will have dissipated long ago.

The Crab pulsar emits optical and X-ray wavelengths in addition to radio waves. Like many others, it produces secondary (weaker) pulses mid-way between the main pulses – simply because its axis of rotation is oriented such that the Earth intercepts both of its rotating beams. Pulsars slow down as they age because the energy they emit in the form of radiation is provided by their rotational kinetic energy. It is because the Crab is a very young pulsar that it is still energetic enough to emit optical and X-ray wavelengths.

5.7 BLACK HOLES

If the core of a collapsing star has a mass that exceeds the upper mass limit of a neutron star (2 to 3 solar masses), it collapses to a point of zero volume and infinite density. There is a region around this point in which the gravitational field is so strong that nothing, not even light, can escape from it. This region is a **black hole** and its radius is called the **Schwarzschild radius**. The boundary (surface) of a black hole is known as its **event horizon**, because we cannot see events that are occurring on the far side of it.

The Schwarzschild radius of a black hole of mass M is given by

$$\text{Schwarzschild radius} = \frac{2GM}{c^2}$$

where G is the gravitational constant ($= 6.67 \times 10^{-11}\,\text{N}\,\text{m}^2\,\text{kg}^{-2}$) and c is the speed of light. A black hole with 3 times the mass of the Sun has a Schwarzschild radius of approximately 9 km.

Black holes are not definitely known to exist, but many astronomers believe that some powerful sources of X-rays provide strong evidence that they do. One such source, **Cygnus X-1**, is known to be a binary system in which a large blue star and an unseen companion are in orbit about their common centre of mass. Calculations show that the mass of this invisible companion is at least 8 times that of the Sun, too big to be a white dwarf or neutron star and therefore possibly a black hole. Gases dragged off the blue star spiral in towards the black hole, and are heated by friction to such high temperatures that they emit X-rays.

QUESTIONS 5A

1. The density, ρ, of a black hole of mass M is given by

$$r = k \times M^n$$

where k and n are constants. Find the value of n.

2. A black hole whose mass is 10 times that of the Sun has a density of $1.85 \times 10^{17}\,\text{kg}\,\text{m}^{-3}$. What is the density of a black hole whose mass is 10^{11} times that of the Sun. (You should not need to perform a detailed calculation.)

5.8 DERIVATION OF THE SCHWARZSCHILD RADIUS

The gravitational force of attraction on a body of mass m at a distance r from a body of mass M is given by Newton's law of gravitation as F where

$$F = G\frac{mM}{r^2}$$

where G is the gravitational constant ($= 6.67 \times 10^{-11}\,\mathrm{N\,m^2\,kg^{-2}}$). The work done δW in moving a small distance δr against this force is given by $W = Fs$ as

$$\delta W = G\frac{mM}{r^2}\delta r$$

Therefore the total work done in moving a body of mass m to infinity from a point at a distance R from a body of mass M is W where

$$W = \int_R^\infty G\frac{mM}{r^2}\,\mathrm{d}r$$

i.e. $$W = GmM\left[\frac{-1}{r}\right]_R^\infty$$

i.e. $$W = G\frac{mM}{R}$$

If the body of mass m is to have sufficient kinetic energy to accomplish this, then its velocity, v, is given by

$$\tfrac{1}{2}mv^2 = G\frac{mM}{R}$$

i.e. $$R = \frac{2GM}{v^2}$$

When $v = c$, R is the Schwarzschild radius and therefore

$$\text{Schwarzschild radius} = \frac{2GM}{c^2}$$

Note Although the approach used here gives the correct result, general relativity is required for a proper derivation of the Schwarzschild radius. We have used Newton's law of gravitation, and this does not apply close to a black hole. Furthermore, we have (effectively) taken the kinetic energy of light to be $\frac{1}{2}mc^2$, which is not correct. One error has compensated for the other.

5.9 THE LENGTH OF TIME SPENT ON THE MAIN SEQUENCE

The main sequence is also a mass sequence – **the greater a star's luminosity, the greater its mass.** Luminosity is a measure of the rate at which a star is using up its hydrogen fuel. It follows that the most massive stars consume fuel more rapidly than smaller stars. Furthermore, despite the fact that they have a greater supply of fuel (fuel supply is proportional to mass), they exhaust it more quickly than less massive stars. Thus, **the more massive a star, the less time it spends on the main sequence.**

A star with the same mass as the Sun probably stays on the main sequence for about 10 billion years. The smallest stars have only 8% the mass of the Sun but will spend a thousand billion years as main-sequence stars. Some of the largest stars, those about 30 times as massive as the Sun, might leave the main sequence after as little as one million years.

QUESTIONS 5B

1. Stars that have 5 times the mass of the Sun are 625 times more luminous. Estimate the time that such a star can be expected to remain on the main sequence given that the Sun is expected to spend 10^{10} years as a main-sequence star.

CONSOLIDATION

Cold clouds of interstellar gas and dust contract under the effect of their own gravity to form protostars. These heat up as gravitational PE is converted to KE, and eventually the temperature is high enough for hydrogen burning to start. Contraction ceases when the sum of the thermal and radiation pressure is equal to the gravitational pressure. Shortly afterwards a temperature equilibrium is established and the protostar is now a stable star on the main sequence.

Stars spend only a tiny fraction of their lives as protostars.

A protostar whose mass is less than 0.08 times that of the Sun becomes a **brown dwarf**.

The energy radiated by a star is generated by **thermonuclear fusion** reactions in its core. The process that occurs in main-sequence stars is known as **hydrogen burning** – the thermonuclear conversion of hydrogen to helium.

A star stays on the main sequence until the hydrogen in its core is exhausted. It then becomes a **red giant**. A red giant whose mass is less than about 4 solar masses eventually becomes a **white dwarf**. A larger red giant throws off its outer layers in a **supernova** explosion; the core is left behind as a **neutron star** or as a **black hole**.

The mass of a **white dwarf** cannot exceed the **Chandrasekhar limit** – 1.4 solar masses.

A pulsar is a rapidly rotating neutron star which emits radio waves that are detected as short pulses of radiation as the beam sweeps past the Earth.

The Schwarzschild radius of a black hole of mass M is given by

$$\text{Schwarzschild radius} = \frac{2GM}{c^2}$$

The greater a star's luminosity, the greater its mass. The greater its mass, the less time it spends on the main sequence.

QUESTIONS ON CHAPTER 5

1. Describe the helium production process by which the energy radiated from a star is generated. Explain why this process takes place only in regions of very high temperature. How do you account for the high temperatures generated within stellar masses before this main energy-releasing process can begin?

[L (specimen)*, '96]

2. The activities which release energy within a main-sequence star such as the Sun cannot begin until the temperature near the star's centre approaches a value of 15 million K.
Explain why this is so and how such a high temperature is produced. [L★, '96]

3. Explain why *black holes* are so named.
A surface for which the escape speed for a body equals the speed of light is called an *event horizon*. Using energy considerations, show from first principles that the radius of the event horizon for a star of mass M is given by

$$R = \frac{2GM}{c^2}$$

where G and c have their usual meanings.
Calculate the radius to which the Sun would have to collapse for it to become a black hole.
Estimate its density at that stage.
($G = 6.67 \times 10^{-11}\,\mathrm{N\,m^2\,kg^{-2}}$, $c = 3.00 \times 10^8\,\mathrm{m\,s^{-1}}$, mass of Sun $= 2.00 \times 10^{30}$ kg.)
[N★, '94]

4. Some stars evolve into *white dwarfs*, others into *red giants*. Explain the terms printed in italic.
Explain briefly what decides the state into which a star evolves when it leaves the main sequence.
[L (specimen)★, '96]

5. It is believed that in its final stages of evolution a star collapses and forms a stable *white dwarf star* or a *neutron star*. State the properties of each type of star. (Do not attempt to explain how they are formed.) [N★, '91]

6. It is thought that black holes are the end states of high-mass main-sequence stars. Explain how, in the right circumstances, it may be possible to detect the presence of a black hole. [L★, '96]

7. Explain each of the following:
(a) All known white dwarf stars are relatively close to the Earth.
(b) The vast majority of stars are main-sequence stars.
(c) There are far more stars whose mass is less than the Sun's than whose mass is greater than the Sun's.

6

THE DOPPLER EFFECT

6.1 THE DOPPLER EFFECT

If an observer and a source of light are in relative motion, the wavelength of the light as measured by the observer is different from the actual wavelength of the light. This is due to a phenomenon known as the **Doppler effect**. When the source and the observer are moving away from each other the shift is to longer wavelengths – the light is said to be **red shifted**. When the source and the observer are approaching each other the shift is to shorter wavelengths, i.e. **blue shifted**. In both cases the extent of the shift increases as the relative velocity of the source and the observer increases.

It can be shown (section 6.6) that the fractional change in wavelength, $\Delta\lambda/\lambda$, is given by

$$\frac{\Delta\lambda}{\lambda} = \frac{v}{c} \quad \text{(for } v << c\text{)} \qquad [6.1]$$

where v is the relative velocity of the source and the observer, and c is the speed of light. The fractional change in frequency, $\Delta f/f$, is given by

$$\frac{\Delta f}{f} = \frac{v}{c} \quad \text{(for } v << c\text{)} \qquad [6.2]$$

Notes (i) Frequency increases when wavelength decreases, and vice versa.

(ii) Equations [6.1] and [6.2] apply to all forms of electromagnetic waves.

6.2 DETERMINING THE SPEEDS OF STARS

When the light emitted by a star (or galaxy) is examined spectroscopically, it is found that each line in the spectrum of any particular element in the star's spectrum occurs at a different wavelength from that of the corresponding line in the laboratory spectrum of the element. The shift in wavelength is taken to be due to the Doppler effect. By measuring the amount by which any particular spectral line is shifted, it is possible to determine the speed of the star from equation [6.1].

Note The Doppler effect cannot be used to find the transverse speed of a star, it gives only the speed of recession or approach, i.e. the speed along the line of sight.

QUESTIONS 6A

1. The wavelengths of the hydrogen H_α and H_β lines when measured using a laboratory source are 6.5628×10^{-7} m and 4.8613×10^{-7}m respectively. The H_α absorption line in the spectrum of a particular star has a wavelength of 6.6560×10^{-7} m. **(a)** Is the distance between the Earth and the star increasing or decreasing? **(b)** What is the speed of the star with respect to the Earth? **(c)** What is the wavelength of the H_β line in the star's spectrum? (The speed of light $= 3.00 \times 10^8\ \mathrm{m\,s^{-1}}$.)

6.3 DETERMINING THE ORBITAL SPEEDS OF BINARY STARS

A binary star consists of two stars in orbit about their common centre of mass. The two stars have the same period of rotation and are always diametrically opposite each other. The ratio of their distances from the centre of mass is equal to the inverse ratio of their masses and therefore unless they happen to have the same mass as each other, they have different orbital radii.

Fig. 6.1 shows such a system, and for simplicity we shall assume that the Earth is in the same plane as the orbits of the stars. At the instant shown, star A is moving directly away from the Earth and B is moving directly towards it. The lines in A's spectrum are therefore red shifted and those in B's are blue shifted. The red shift is larger than the blue shift because A's orbital velocity is larger than B's. As the stars continue to move along their orbits, the two sets of spectral lines will move closer together. After a quarter of a revolution, B will eclipse A and therefore A's spectrum will temporarily disappear. (If we could see A's spectrum, it would coincide with B's at this stage.) For the next quarter of an orbit, the two sets of spectral lines become progressively farther apart, A's displaced towards the blue and B's towards the red. The lines then move closer together again until A eclipses B. Thus the two sets of spectral lines repeatedly move back and forth across each other. If the maximum shift in each set of lines is measured, the orbital speeds of A and B can be found from equation [6.1].

Fig. 6.1
A binary star

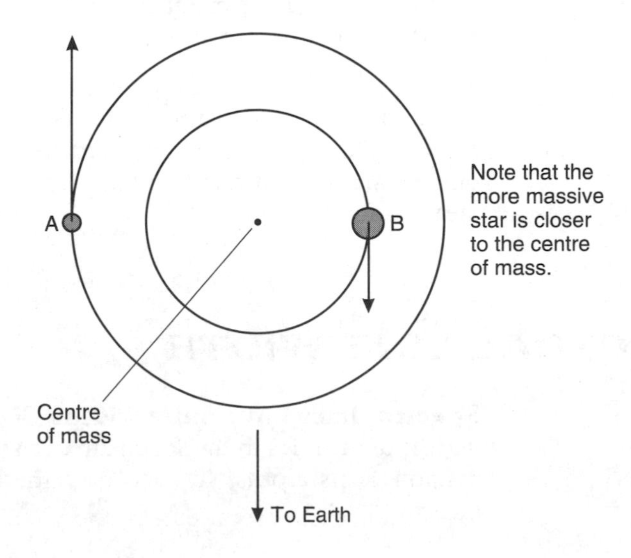

Spectra of the binary system zeta UMa when the two component stars are moving (a) perpendicular to the line of sight, (b) parallel to the line of sight and therefore exhibiting line doubling.

Note If there is any Doppler shift in the lines when the spectra coincide (i.e. when one star is eclipsing the other), it is because the system as a whole is moving relative to the Earth.

QUESTIONS 6B

1. The wavelength of the H_α line in the spectrum of one of the stars of a binary system varies from a minimum of 6.5562×10^{-7} m to a maximum of 6.5694×10^{-7}m, and does so over an interval of 100 days. The wavelength of the hydrogen H_α line when measured using a laboratory source is 6.5628×10^{-7} m. Find **(a)** the speed of the star, **(b)** the radius of its orbit. You may assume that the observations have been made from a point in the plane of the orbit and that the orbit is circular. (The speed of light = 3.00×10^8 m s^{-1}.)

2. When the binary system in Question 1 is observed from some distant planetary system the wavelength of the H_α line varies from 6.5662×10^{-7} m to 6.5794×10^{-7} m. What is the speed of the centre of mass of the binary with respect to this observer?

6.4 MEASURING THE ROTATION OF THE SUN AND PLANETS

The Doppler effect has been used to find the speed at which the Sun is rotating. The **Fraunhofer lines** (absorption lines) originating at that side of the Sun's disc which is approaching us are shifted towards the blue, those from the other side of the disc are shifted towards the red. The extent to which either of these sets of lines is shifted from those produced at the centre of the disc allows us to calculate the speed of the Sun's rotation.

The rotational speed of a planet (or a planetary ring system) can be found by comparing the spectrum of the Sun's light reflected from one edge of the planetary disc with that reflected from the centre of it. Rotational speeds of planets are also found from the Doppler shift of radar waves reflected off them (see section 9.4).

6.5 SPECTRAL LINE WIDTH

Spectral lines have finite widths. Two important factors that determine the widths of the lines in the spectrum of a star are the star's rotation and the thermal motion of its atoms. In each case the line broadening is due to the Doppler effect.

Rotational broadening

If a star is rotating, the light that reaches us from one side of it will be red shifted and that from the other side will be blue shifted. Furthermore, light coming from the edges of the star will be shifted more than that from nearer the middle. We therefore receive a whole range of wavelengths, centred on that of the light coming from the centre of the star's disc.

Thermal broadening

This is caused by the thermal motion of the atoms. The light emitted by an atom that is moving away from us is red shifted, that from an atom that is moving towards us is blue shifted. Since the atoms in a star are moving around at random, we receive a whole range of wavelengths and the spectral lines are therefore broader than they would otherwise be.

At any temperature T

$$\text{The average KE } \left(\tfrac{1}{2}mv^2\right) \text{ of an atom } = \tfrac{3}{2}kT$$

where k is Boltzmann's constant. It follows that at any given temperature the least massive atoms have the highest average speeds, and therefore the **spectral lines of light elements exhibit larger thermal broadening than those of heavier elements.**

6.6 DERIVATION OF EQUATIONS [6.1] AND [6.2]

The derivation that follows is valid only when $v << c$. If v is comparable with c, a relativistic treatment is necessary.

Consider a star moving directly away from the Earth with velocity v relative to the Earth, emitting light of wavelength λ and speed c. The time interval between the star emitting two successive crests (i.e. the time for it to emit one wavelength) is λ/c. By the time the second crest is emitted, the first crest will have moved towards the Earth by λ and the star will have moved away by $v \times (\lambda/c)$. Therefore

$$\text{Separation of two successive crests } = \lambda + v \times \frac{\lambda}{c}$$

i.e. $$\text{Wavelength measured on Earth } = \lambda + v \times \frac{\lambda}{c}$$

The change in wavelength, $\Delta\lambda$, is therefore given by

$$\Delta\lambda = v \times \frac{\lambda}{c}$$

i.e. $$\frac{\Delta\lambda}{\lambda} = \frac{v}{c}$$

Since $v << c$, the fractional change in wavelength is very nearly the same as the fractional change in frequency and therefore we may put

$$\frac{\Delta f}{f} = \frac{v}{c}$$

QUESTIONS ON CHAPTER 6

1.

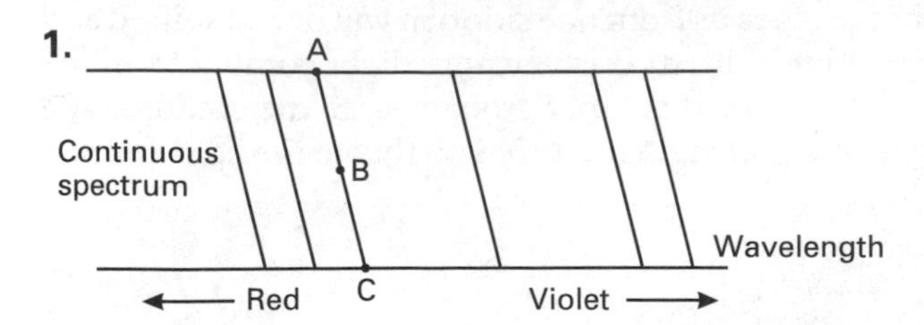

The figure shows part of the spectrum of light received from a planet. It consists of dark lines running from top to bottom over a bright horizontal continuous spectrum.

(a) Explain the origin of the continuous spectrum and the origin of the dark lines.

(b) Explain why the dark lines slope.

(c) If the wavelengths at points A, B and C on the dark line are 5.00017×10^{-7} m, 5.00000×10^{-7} m and 4.99983×10^{-7} m respectively, calculate the period of rotation of the planet, given that the radius of the planet is 6.0×10^7 m.
(Speed of light $= 3.0 \times 10^8$ m s^{-1}.)

[N*, '91]

2. (a) Give **two** reasons why spectral lines of a star have a finite width. Explain qualitatively how, in each case, the finite width is produced.

(b) The helium yellow line of wavelength 5.88×10^{-7} m observed in the spectrum of a star has a spread of wavelength of 6.70×10^{-11} m. Calculate the maximum thermal velocity of helium atoms in the photosphere of this star.

(Speed of light in vacuo $= 3.00 \times 10^8$ m s^{-1}.)

[N]

3. The helium yellow line of wavelength 5.88×10^{-7} m in the spectrum of a star exhibits a maximum spread of wavelength, due to *thermal Doppler broadening*, of 4.30×10^{-11} m.

(a) Explain the term in italics and indicate, qualitatively, how this effect is caused.

(b) Estimate the maximum random thermal velocity of helium atoms in the photosphere of the star.
The speed of light is 3.00×10^8 m s^{-1}.

(c) Describe and account for the appearance of the spectral lines produced by a rapidly spinning star.

[N]

4. The apparent magnitude of the binary star system β Persei and the wavelength of the calcium K line in its spectrum are measured at intervals of time.
Figure 1 shows how the apparent magnitude of the star system varies with time and Figure 2 shows how the wavelength of the calcium K line varies with time.

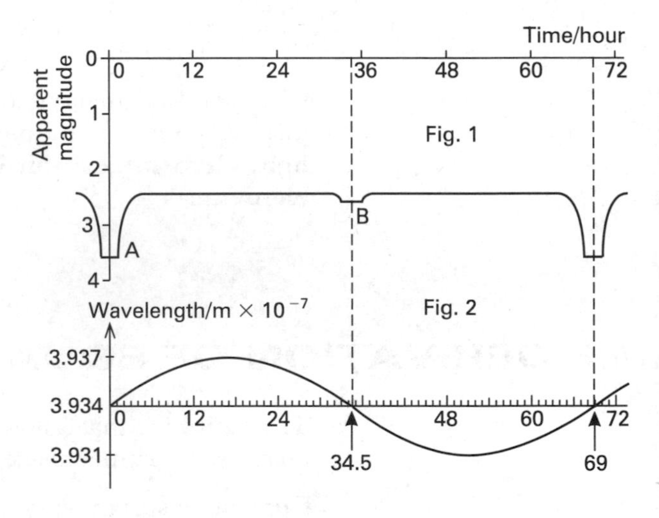

(a) Explain why the magnitude of the star system varies in the manner shown in Fig. 1. Why is the apparent magnitude at A different from that at B?

(b) (i) Explain why the wavelength varies in a sinusoidal manner as shown in Fig. 2.

(ii) Calculate the orbital speed of the brighter star, stating the assumption which must be made.

(iii) Deduce the period of revolution of this star and hence calculate the radius of its circular orbit.
The speed of light in vacuo $= 3.00 \times 10^8$ m s^{-1}.

[N]

7

THE EXPANDING UNIVERSE

7.1 HUBBLE'S LAW

Edwin Hubble (1889–1953) seen in front of the 2.5 m telescope at the Mount Wilson Observatory

Nearly all galaxies have absorption spectra that are red shifted. If these shifts arc taken to be due to the Doppler effect, it means that nearly all the galaxies are moving away from the Earth. (The only exceptions are some of the nearby galaxies.) In 1929 an American astronomer, Edwin Hubble, discovered that

> Except for a few nearby galaxies, the speed at which a galaxy is receding from us is proportional to its distance from us (**Hubble's law**).

This should not be taken to imply that we occupy some unique place in the Universe – the galaxies are not only moving away from us, they are also moving away from each other. If Hubble had been able to make his measurements from any other galaxy, he would have obtained an equivalent result. This is entirely consistent with the **cosmological principle**, which assumes that, on a large scale, any part of the Universe is essentially the same as every other part.

Note The nearby galaxies do not conform to Hubble's law because the speed at which they would be expected to recede is low and is masked by their random motions within their local groups.

7.2 THE BIG BANG

According to Hubble's law, the galaxies are receding at speeds proportional to their distances from us, i.e. those that are farthest away are moving the fastest. The most commonly accepted explanation of this is that at some time in the past the Universe was in a state of extremely high concentration, and that as a result of some gigantic explosion (the **Big Bang**) it has been expanding ever since.

Note This does not mean that all the matter in the Universe was once concentrated at a single point in the Universe, it means that that point was the Universe. The Big Bang created an expansion of space itself; it did not cause matter to rush outwards to occupy an already existing space. This is not an easy concept; it may help to consider two points, A and B, on the surface of a balloon (Fig. 7.1(a)). If we inflate the balloon (Fig. 7.1(b)), A and B become farther apart because the space they occupy, the surface of the balloon, has expanded. They have not moved apart by moving through an already existing space as in Fig. 7.1(c).

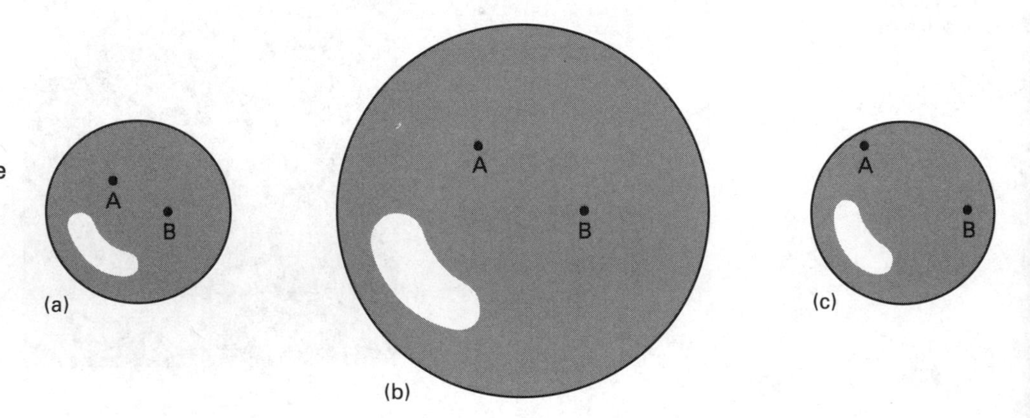

Fig. 7.1
To illustrate the difference between the expansion of space and matter moving through an already existing space

It is also worth pointing out that, just as the surface of the balloon has no edge and no centre, it is meaningless to think of the Universe as having an edge or a centre.

7.3 THE HUBBLE CONSTANT AND THE AGE OF THE UNIVERSE

Hubble's law can be expressed as

$$v = H_0 d \qquad [7.1]$$

where v is the velocity at which a galaxy a distance d away from us is receding, and H_0 is a constant known as **Hubble's constant** (Fig. 7.2). The value of the Hubble constant is believed to lie somewhere in the range 50 to 100 km s^{-1} Mpc^{-1}. The uncertainty is primarily due to the difficulty in measuring the distances of the more distant galaxies. Fig. 7.2 has been drawn on the assumption that $H_0 = 75$ km s^{-1} Mpc^{-1}, i.e. that the speed of recession of a galaxy 1 megaparsec (1×10^6 pc) away is 75 km s^{-1}.

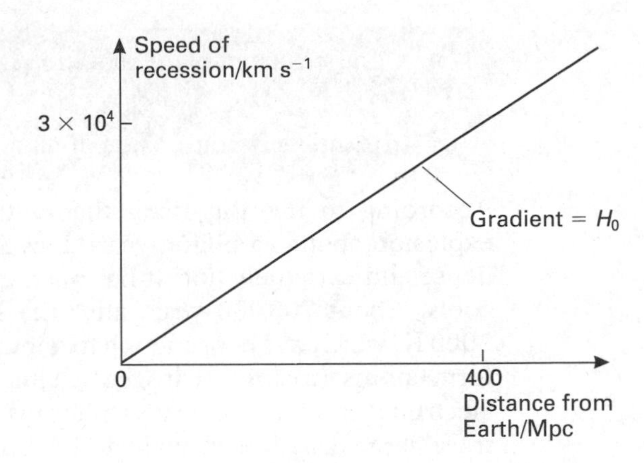

Fig. 7.2 Hubble's law

It follows from equation [7.1] that matter a distance d away from us has been travelling for a time t where

$$t = \frac{d}{v} = \frac{1}{H_0}$$

Since this matter is supposed to have been moving since the Universe began,

$$\text{Age of Universe} = \frac{1}{H_0} \qquad [7.2]$$

We can expect the actual age to be somewhat less than this because the galaxies will have been slowing down as a result of their mutual gravitational attraction. When this and the uncertainty in the value of H_0 are taken into account, the age of the Universe is estimated to be between 7 and 20 billion years. Many astronomers favour a figure of 15 billion (1.5×10^{10}) years.

Note It is conventional to express H_0 in km s^{-1} Mpc^{-1}. Since 1 Mpc = 3.1×10^{19} km, it follows that

$$H_0 \text{ in s}^{-1} = H_0 \text{ in km s}^{-1} \text{ Mpc}^{-1} \div 3.1 \times 10^{19}$$

QUESTIONS 7A

1. Spectroscopic investigations of a distant galaxy reveal that it is receding from the Earth at 7.4×10^7 m s^{-1}. Use Hubble's law to determine the distance of the galaxy from the Earth **(a)** in metres, **(b)** in light-years. (Hubble's constant = 2.4×10^{-18} s^{-1}, speed of light = 3.0×10^8 m s^{-1}.)

2. How far from the Earth (in Mpc) is a galaxy that has a recessional velocity of 6.0×10^3 km s^{-1} if $H_0 = 75$ km s^{-1} Mpc^{-1}?

3. Estimate the age of the Universe, in years, on the basis of $H_0 = 75$ km s^{-1} Mpc^{-1}. (1 Mpc = 3.1×10^{19} km.)

7.4 THE COSMIC MICROWAVE BACKGROUND RADIATION

This is electromagnetic radiation that reaches the Earth from all parts of the sky. It was first detected by Arno Penzias and Robert Wilson in 1965 and is regarded as strong evidence in support of the Big Bang theory.

(i) The spectrum of the radiation is that of a black body at a temperature of 2.73 K.

(ii) Its intensity is the same in all directions (to about 1 part in 10^5).

According to the Big Bang theory the Universe originated in an immense explosion about 15 billion years ago. At that stage the Universe was extremely dense and extremely hot. It has been expanding ever since and as it expands, it cools. About 700 000 years after the Big Bang the temperature had fallen to 3000 K, which was cool enough to allow electrons to become attached to nuclei to form atoms (mainly hydrogen). This drastically reduced the number of free electrons* available to scatter photons and therefore photons were now free to travel large distances virtually unimpeded by matter – the Universe had become transparent for the first time. These are the photons that make up the microwave background radiation. The Universe has expanded by a factor of about 1000 since it became transparent. The wavelength of any radiation that has survived since that time will also have increased by a factor of 1000. It follows from Wien's law (section 3.2) that the temperature should have fallen from 3000 K to 3 K, which is in excellent agreement with the observed value of 2.73 K.

7.5 EVIDENCE SUPPORTING THE BIG BANG MODEL

The Big Bang model accounts for the expansion of the Universe. It has had two other notable successes.

(i) It predicted the existence of the microwave background radiation discovered in 1965 (Section 7.4.).

(ii) It predicts that the ratio, by mass, of hydrogen to helium should be 3 to 1 and this has been confirmed by observation.

It is these two successes, rather than its ability to account for the expansion, that has made the Big Bang model the most widely accepted theory of the Universe. The idea that the Universe is expanding is based on Hubble's law, which is itself based on the belief that the galactic red shifts are due to the Doppler effect. Some cosmologists think the red shifts are due to a gravitational effect rather than to the Doppler effect. If they are correct and the Big Bang model did no more than account for the expansion, it is unlikely that it would have the widespread acceptance that it has.

7.6 THE EVOLUTION OF THE UNIVERSE

Photons and a variety of fundamental particles (e.g. quarks, electrons, neutrinos and their antiparticles) were created within a tiny fraction of a second of the Big Bang.

As the Universe cooled and average photon and particle energies decreased, quarks combined to form neutrons and protons, which subsequently combined to form nuclei. Later, nuclei joined with electrons to form atoms, and finally, atoms were pulled together by gravity to form stars and galaxies.

*Free electrons scatter photons far more effectively than atoms do.

The stage at which any particular type of particle comes into existence depends on the binding energy of the particle concerned. The higher the binding energy, the higher the temperature at which the particle can form without breaking up again. Atomic hydrogen, for example, could not exist whilst the temperature was so high that a significant number of particles had enough energy to ionize it.

A very much simplified account of how the Universe is thought to have evolved after the first 10^{-6} s is given in Table 7.1.

Table 7.1
The evolution of the Universe

Time since Big Bang,	*Temperature*	*Main features*
10^{-6} s	10^{13} K	Quarks combine to form nucleons (protons and neutrons) and **antinucleons**.
10^{-2} s	10^{11} K	Nucleon–antinucleon annihilation leaves a **large excess of nucleons over antinucleons** (section 7.7).
14 s	3×10^{9} K	Electron–positron annihilation leaves a **large excess of electrons over positrons**.
100 s	10^{9} K	Neutrons and protons fuse to form **helium nuclei** (Note (i)). Small amounts of ^{7}Li and ^{7}Be are also produced, after which the building of nuclei very nearly ceases because there are no stable nuclei with mass numbers of either 5 or 8. The Universe is now a high-temperature **plasma** consisting mainly of photons, protons, helium nuclei, electrons and neutrinos.
7×10^{5} y	3000 K	Electrons become attached to nuclei to form **atoms** (mainly hydrogen and helium). Drastic reduction in number of free electrons allows photons to travel freely, i.e. Universe becomes transparent to photons (section 7.4). No nuclei–nuclei repulsion now that atoms have formed and therefore **galaxies** and **stars** form under the effect of gravity (Note (ii)). Thermonuclear fusion occurs in stars, initially producing ^{4}He and then nuclides of higher mass number up to ^{56}Fe (sections 5.2 and 5.4). Heavier nuclei are produced in stars by neutron capture followed by β^{-} decay, and by (endothermic) fusion reactions in supernovae explosions (section 5.4).

Notes

(i) There are two processes, one involving deuterium (^{2}H) and tritium (^{3}H), the other involving deuterium and helium 3 (^{3}He).

$$p + n \rightarrow {}^{2}H \quad \text{then} \quad {}^{2}H + n \rightarrow {}^{3}H \quad \text{then} \quad {}^{3}H + p \rightarrow {}^{4}He$$

and

$$p + n \rightarrow {}^{2}H \quad \text{then} \quad {}^{2}H + p \rightarrow {}^{3}He \quad \text{then} \quad {}^{3}He + n \rightarrow {}^{4}He$$

(ii) The COBE* satellite detected slight (1 part in 10^{5}) variations (**ripples**) in the temperature of the microwave background radiation in 1992. This implies that there were correspondingly slight variations in the density of the Universe when it first became transparent to radiation. Gravitational attraction would have drawn material (atoms of hydrogen and helium) towards any region that was a little more dense than those around it, amplifying the density variations (**lumpiness**) and eventually forming stars, galaxies and clusters of galaxies.

(iii) Most of the experimental evidence to support the evolution of the Universe as outlined in Table 7.1 comes from experiments in **particle physics**. Fundamental particles are accelerated to very high energies and are then made to smash into each other to produce particles that do not exist in the

*COBE is an acronym for Cosmic Background Explorer.

Universe of today. Since temperature is a measure of kinetic energy, high-energy experiments are effectively high-temperature experiments. Energies corresponding to temperatures as high as 10^{15} K can be achieved with proton synchrotrons, providing us with evidence of the reactions that might be expected to have been taking place in the Universe a mere 10^{-10} s after the Big Bang.

7.7 MATTER AND ANTIMATTER

In the very early Universe there was an equilibrium between the rate at which photons produced particle–antiparticle pairs (in a process known as **pair-production**) and that at which particle–antiparticle pairs **annihilated** each other to produce photons. There is a threshold energy below which a photon cannot produce a particular particle–antiparticle pair. As the Universe cooled and photon energies decreased, pair production started to occur less frequently than annihilation, and so the ratio of particles (and antiparticles) to photons started to fall. Nucleon–antinucleon production ceased when the temperature had fallen to about 10^{11} K. Electron–positron production was able to continue until the temperature had fallen to about 3×10^{9} K. (Electrons and positrons have less mass than nucleons and antinucleons and therefore it takes less energy to create an electron–positron pair than it does to create a nucleon–antinucleon pair.)

If particles and their antiparticles had been created in (exactly) equal numbers, all the matter and antimatter would have been annihilated and the Universe would now contain nothing but photons. It is thought that some symmetry-breaking mechanism existed in the earliest moments of the Universe that allowed the number of particles to exceed the number of antiparticles by one part in a billion. For every billion antiprotons, for example, there would have been a billion and one protons, and therefore one proton in every billion would escape annihilation. The end result is a Universe in which photons outnumber protons by a factor of a billion, and which is composed of matter rather than antimatter – exactly what is observed.

7.8 THE FUTURE OF THE UNIVERSE

The rate at which the Universe is expanding is slowing down because of the mutual gravitational attraction of the galaxies. If this attraction is strong enough, the expansion will eventually stop and the Universe will start to contract, ending in what is known as **the Big Crunch**. Such a Universe is said to be **closed** or **bounded**; one which never stops expanding is **open** or **unbounded**. A Universe in which the rate of expansion would become exactly zero after an infinite time is said to be **flat** or **marginally bounded**.

Whether the Universe is open, closed or flat depends on how its average density compares with a value known as the **critical density** (ρ_c). The three possibilities are shown graphically in Fig. 7.4. We do not know which of the three curves we are on. If we are on the bottom curve, we must be left of centre because we know that the Universe is expanding at the present time.

Fig. 7.4
Open, flat and closed Universes compared

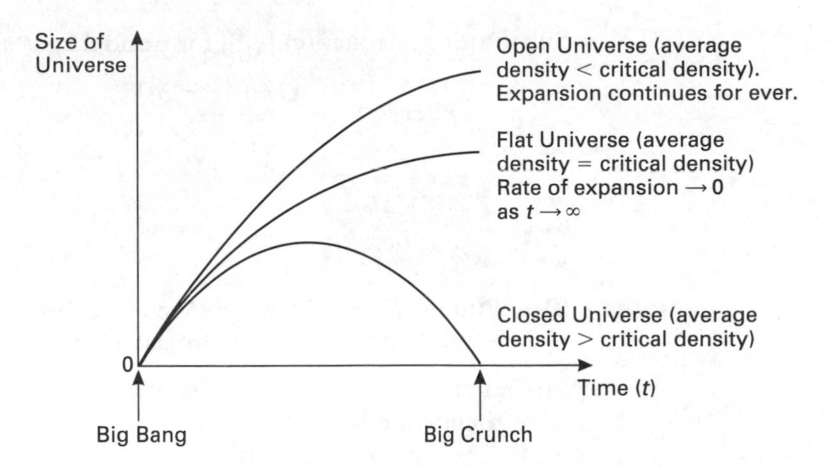

There is some uncertainty about the value of ρ_c because it depends on the value of H_0, which is itself uncertain. The value of the average density is even more uncertain. Estimates based on the total mass of visible material in the Universe suggest that it is only a few per cent of the critical density. This implies that the Universe is open and will expand for ever. However, it appears from observations of the motions of galactic clusters that the Universe contains a large amount of **dark matter**, non-luminous matter that is otherwise undetectable. Taking this dark matter into account gives a much higher estimate of the average density. This, and a number of other considerations, suggest that the average density is very nearly (and possibly exactly) equal to the critical density.

An accurate value for the density of the Universe would allow us to make a much better estimate of its age than can be obtained from equation [7.2] alone, because it would allow us to work out the rate at which the expansion has been slowing down.

7.9 HUBBLE'S CONSTANT AND CRITICAL DENSITY

Consider a spherical region of the Universe of mass M and radius r centred on the site of the Big Bang. Consider also a galaxy of mass m and velocity v that is just outside this region. The net gravitational force on the galaxy is due only to the spherical region (because the distribution of the rest of the matter in the Universe can be assumed to have spherical symmetry on the basis of the cosmological principle) and therefore

$$\text{Gravitational PE} = -G\frac{mM}{r}$$

If the galaxy has just sufficient kinetic energy to reach infinity (where its gravitational PE would be zero),

$$\tfrac{1}{2}mv^2 = G\frac{mM}{r} \qquad [7.3]$$

By Hubble's law

$$v = H_0 r$$

If the galaxy just reaches infinity, the density of the Universe is the critical density, ρ_c, and therefore

$$M = \tfrac{4}{3}\pi r^3 \rho_c$$

Substituting in equation [7.3] for v and M gives

$$\tfrac{1}{2}m(H_0 r)^2 = \frac{Gm(\tfrac{4}{3}\pi r^3 \rho_c)}{r}$$

i.e.

$$\rho_c = \frac{3H_0^2}{8\pi G}$$

Notes (i) Putting $H_0 = 75\,\text{km s}^{-1}\,\text{Mpc}^{-1}$ gives $\rho_c = 1.0 \times 10^{-26}\,\text{kg m}^{-3}$, which is equivalent to about 6 hydrogen atoms per cubic metre.

(ii) A proper derivation of the expression for critical density requires the general theory of relativity.

7.10 OLBERS' PARADOX

It used to be thought that the Universe was infinite, static and uniformly populated with stars. If it is, then there is bound to be a star somewhere along every line of sight from the Earth, and therefore the night sky should be infinitely bright. (The dimness of the more distant stars is made up for by the fact that there are more of them.) This creates a dilemma, because the night sky is not infinitely bright – it is known as **Olbers' paradox,** after the German astronomer who brought it to public attention in 1826.

Olbers reached a false conclusion by sound reasoning, and therefore one or more of the underlying postulates must be invalid. We now know, of course, that two of them are invalid – the Universe is neither infinite nor static, and this provides us with two ways of resolving the paradox.

(i) Hubble's law implies that the age of the Universe is finite, and therefore there may be some stars that are so far away that their light has not yet had time to reach us. (This would be true of an even greater proportion of the stars if the Universe were infinite in extent.)

(ii) We know from Hubble's law that the light that reaches us from distant galaxies is red shifted and that the most distant galaxies have the greatest red shifts. Some galaxies are so far away that all of their visible light has been shifted into the infrared and therefore cannot be seen.

Note The reader may feel that Olbers' paradox could have been resolved simply by rejecting the assumption of an infinite Universe. It could not – the Universe was believed to be static and therefore had to be infinite to prevent the stars all crowding together under the effect of their mutual gravity.

7.11 QUASARS

Quasars are **extremely luminous** objects with **very large red shifts.**

On the basis of Hubble's law, the red shift values indicate that quasars are the most distant objects known – typically of the order of 10 billion (10^{10}) light-years away from the Earth. Because they are so distant, they allow us to see back to a time when the Universe was in its infancy.

The very fact that we can see them at these great distances (albeit with powerful telescopes) means that quasars must be extremely bright objects. **A typical quasar is a thousand times brighter than an entire galaxy**. They vary in brightness, with periods ranging from a few days to a few months. This implies that the smallest of them can be no more than a few light-days in diameter, because the brightness of an object cannot vary more quickly than it takes light to travel from one side of it to the other (see Fig. 7.5). The diameter of a typical galaxy can be over a million times that of a quasar, yet the quasar is a thousand times brighter. The source of this enormous amount of energy is thought be the gravitational energy lost by matter that is being sucked into a massive black hole at the centre of the quasar. A quasar of average luminosity would require a black hole with the mass of a billion Suns!

Fig. 7.5
To illustrate how the maximum size of a quasar is deduced

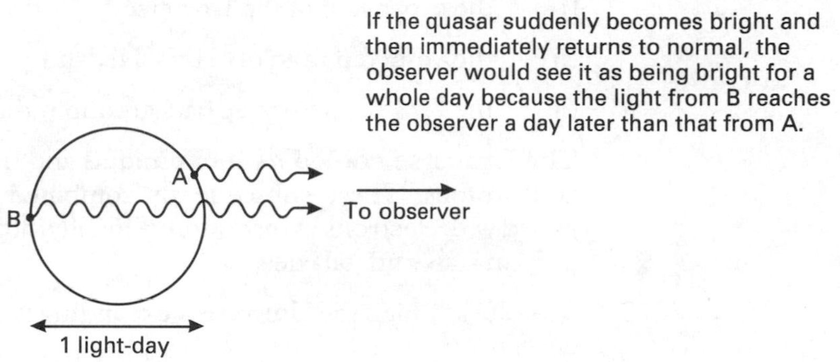

If the quasar suddenly becomes bright and then immediately returns to normal, the observer would see it as being bright for a whole day because the light from B reaches the observer a day later than that from A.

A
B
To observer
1 light-day

Notes (i) Quasar is an abbreviation of 'quasi-stellar radio source'. The name arose because the first quasars to be discovered had the appearance of stars but, unlike ordinary stars, were strong radio emitters. It is now known that only about 10% of quasars are strong sources of radio waves but the name is still used.

(ii) The spectra of quasars show strong emission lines superimposed on a continuous background; those of most stars and galaxies exhibit absorption lines.

QUESTIONS 7B

1. Sirius, the brightest star in the sky, has a luminosity of 8.9×10^{27} W and is 8.7 light-years away. Estimate the luminosity of a quasar which is at a distance of 3.0×10^9 light-years if the apparent brightness of Sirius is 2.0×10^5 times that of the quasar.

CONSOLIDATION

Nearly all galaxies have absorption spectra that are red shifted. If this is due to the Doppler effect, the galaxies must be moving away from us (and from each other).

Hubble's law

The speed (v) at which a galaxy is receding from us is proportional to its distance (d) from us, i.e.

$$v = H_0 d$$

where H_0 is **Hubble's constant.**

Hubble's law implies that the Universe was once in a state of extremely high concentration and that it has been expanding ever since.

$$\text{Age of Universe} = \frac{1}{H_0}$$

The Big Bang model accounts for

(i) the expansion of the Universe,

(ii) the observed ratio of ^{1}H to ^{4}He, and

(iii) the 2.73 K microwave background radiation.

The Universe cooled as it expanded and quarks combined to form neutrons and protons. These subsequently combined to form nuclei which eventually joined with electrons to form atoms. Finally, atoms were pulled together by gravity to form stars and galaxies.

The rate at which the Universe is expanding is decreasing because of gravitational effects.

If the average density of the Universe is less than the **critical density** (ρ_c), the Universe will expand for ever (i.e. an **open Universe**). If the average density is greater than the critical density, the Universe will eventually contract (i.e. a **closed Universe**). If the average density is equal to the critical density, the rate of expansion will become zero after an infinite time (i.e. a **flat Universe**).

$$\rho_c = \frac{3H_0^2}{8\pi G}$$

QUESTIONS ON CHAPTER 7

1. Explain the importance of the Doppler effect in providing evidence for the Big Bang model of the Universe. State Hubble's law and from it deduce the units of the Hubble constant, H. Current estimates of H vary by as much as a factor of 2. What is the problem in obtaining a reliable value of H?
The approximate age of the Universe, t_0, can be obtained from the formula $t_0 = 1/H$. What assumption has been made in obtaining this formula? [L*, '95]

2. **(a)** In a laboratory source the calcium H and K lines have wavelengths of 3.968×10^{-7} m and 3.934×10^{-7} m, respectively. When the same lines are measured in the absorption spectrum of a faint galaxy in the constellation Hydra, the wavelength of the H line is found to be greater by 0.806×10^{-7} m than that measured on Earth.

(i) Explain why the wavelength measured in the spectrum of the galaxy is greater than that measured on Earth.

(ii) Explain why the increase in the wavelength of the H line should be different from the increase in the wavelength of the K line. Calculate the wavelength of the K line when measured in the spectrum of the galaxy.

(iii) Calculate the velocity of the galaxy in Hydra with respect to the Earth.

(iv) If the Hubble constant is taken to be 100 km s^{-1} Mpc^{-1}, calculate, in light-years, the distance of the galaxy from Earth.

(b) There is large uncertainty in the value of the Hubble constant. If the distance of another galaxy from Earth is known, explain how the distance of Hydra could be obtained without using the Hubble constant.

(c) The galaxies at the 'edge' of an expanding Universe have the greatest speed.

(i) Estimate the maximum size of the Universe if the Earth is taken to be near the centre and if the Hubble constant is $3.24 \times 10^{-18}\ s^{-1}$.

(ii) Estimate also the age of the Universe, assuming that it is expanding at a constant rate.

(Speed of light = $3.00 \times 10^8\ m\,s^{-1}$, 1 pc = 3.26 light-years.) [N, '92]

3. What evidence is there **(a)** that the Universe is expanding, **(b)** that it began as a Big Bang?

4. The Universe is said to be expanding. The diagram shows a deflated balloon with three points A, B and C marked on it.

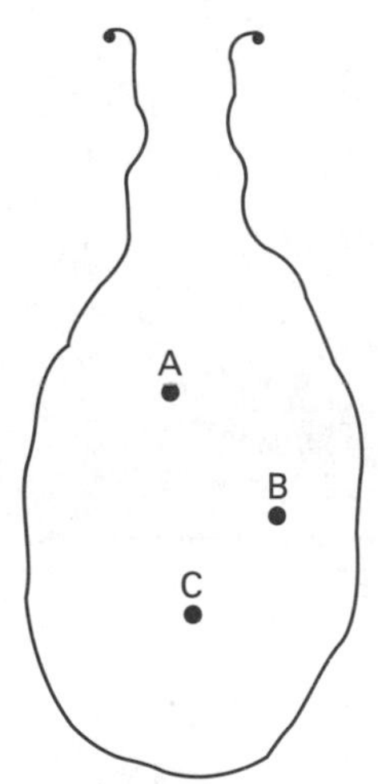

Draw labelled diagrams showing what the balloon would look like when it is

(a) half inflated, and

(b) fully inflated,

and use your diagrams to explain the concept of an expanding universe. [L*, '91]

5. Explain briefly how studies of line spectra from distant galaxies lead to the conclusion that the Universe is expanding.

Cosmologists often use diagrams similar to that below when describing the expansion of the Universe.

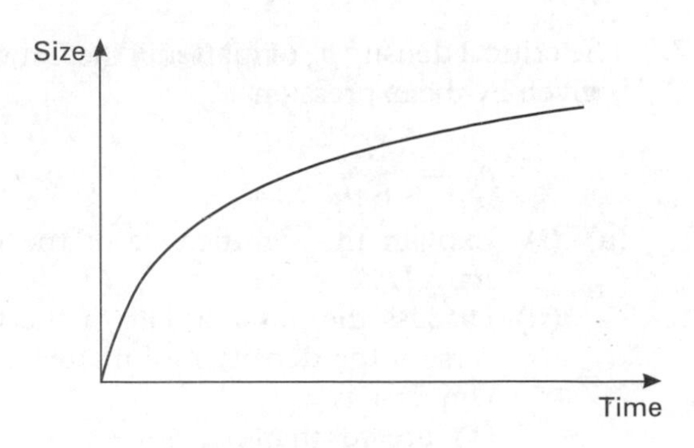

In two or three sentences, describe the way in which the expansion rate changes as the Universe evolves and state the reason for this change.

In the diagram above, the Universe is assumed to have an average mass density exactly equal to the 'critical density'.

The matter so far observed in the Universe suggests an average density below the critical value. Copy the above diagram and on it show how the Universe would evolve if this were the case.

It is possible that the neutrino has a finite small mass. It is thought that neutrinos are very abundant in the Universe, and that their mass may be sufficient to raise the average density above the critical value. On your copy of the diagram, show how the Universe would evolve in this case.

On your diagram label the 'Big Bang' and the 'Big Crunch'. [L (specimen), '92]

6. In the absorption spectrum of a particular faint galaxy the wavelength of the calcium K line is found to be greater than the wavelength of the same line when produced by a laboratory source.

(a) Explain why there is this increase in wavelength.

(b) Explain how this 'red-shift' in the wavelength can be used to measure the distance of the galaxy from Earth, provided the Hubble constant is known.

(c) If the two measured values of the wavelength are 3.934×10^{-7} m and 4.733×10^{-7} m, calculate the speed of the galaxy with respect to the Earth.

(d) Hence calculate the distance in light-years of the galaxy from Earth.

Take the Hubble constant to be $5.0 \times 10^4\ m\,s^{-1}\,Mpc^{-1}$, 1 pc = 3.26 light-years and the speed of light to be $3.00 \times 10^8\ m\,s^{-1}$. [N, '96]

7. The critical density ρ_0 of matter in the Universe is given by the expression

$$\rho_0 = \frac{3H_0^2}{8\pi G}$$

(a) (i) Explain the significance of the constant H_0.

(ii) Discuss the possible fate of the Universe if the density ρ of matter in the Universe is

(1) greater than ρ_0,

(2) less than ρ_0.

(b) The diagram shows possible curves for the variation with time of the size of the Universe.

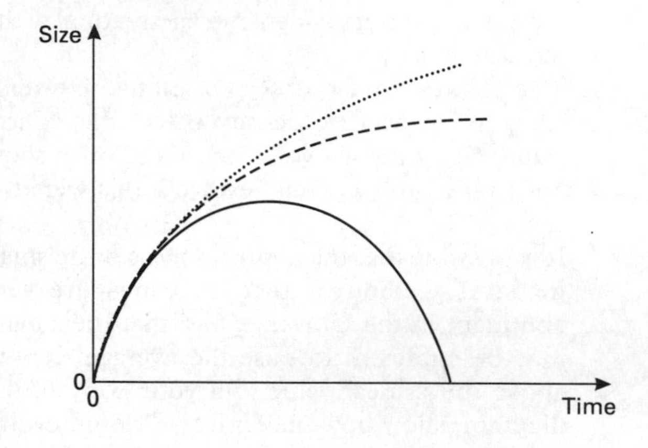

Copy the diagram and on it

(i) label the curve corresponding to $\rho > \rho_0$,

(ii) label the curve corresponding to $\rho < \rho_0$,

(iii) mark on the diagram a point to represent the Universe at the present time.

(c) Discuss your answer to **(b) (iii)** with reference to the possible fate of the Universe. [C*, '96]

8

OPTICAL ASTRONOMY

The reader should be familiar with ray diagrams, the properties of lenses and curved mirrors, and with the lens and mirror formulae before continuing* .

8.1 ANGULAR MAGNIFICATION

The size that an object appears to be is determined by the size, b, of its image on the retina (Fig. 8.1). For small angles

$$b = a\theta$$

where a is the length of the eyeball and θ is the angle subtended at the eye by the object (i.e. the **visual angle**). Since a is constant, **the size of the image on the retina (and therefore the apparent size of the object) is proportional to the visual angle**. For example, a penny held at arm's length subtends a larger angle at the eye than the Moon does, and therefore can block it from view. Thus, although the penny is smaller than the Moon, it appears to be bigger because it is subtending a larger angle at the eye.

Fig. 8.1
Image formation by the eye

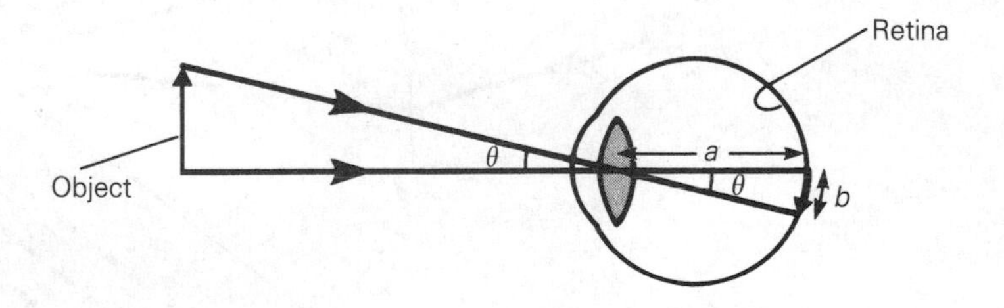

The angular magnification (also known as **magnifying power**), M, of an optical instrument is defined by

$$M = \frac{\beta}{\alpha} \qquad [8.1]$$

where

β = the angle subtended at the eye by the image

α = the angle subtended at the unaided eye by the object.

8.2 THE ASTRONOMICAL (REFRACTING) TELESCOPE

This consists of two converging lenses. The **objective lens** produces a real image of the object being viewed. This (intermediate) image acts as an object for the

*See, for example, R. Muncaster, *A-level Physics* – sections 19.1–19.4 and 20.4–20.7.

eyepiece lens which, behaving as a magnifying glass, produces a virtual image of it. If the magnifying power of the instrument is to be high, the focal length of the objective lens must be large and that of the eyepiece must be small.

Telescopes are used to view objects which are at great distances, and in each of the situations discussed here it will be assumed that the object is at infinity. Because the diameter of the objective lens is small compared with the distance of the object from the lens, all the rays reaching the lens from a single point on the object can be taken to be parallel to each other.

Final Image at Infinity (i.e. Normal Adjustment)

The arrangement is shown in Fig. 8.2. The object is at infinity, and therefore the intermediate image is in the focal plane of the objective lens. The separation of the lenses is such that their focal planes coincide, and therefore the eyepiece lens, acting as a magnifying glass, produces a final image which is at infinity. The eye is relaxed (unaccommodated).

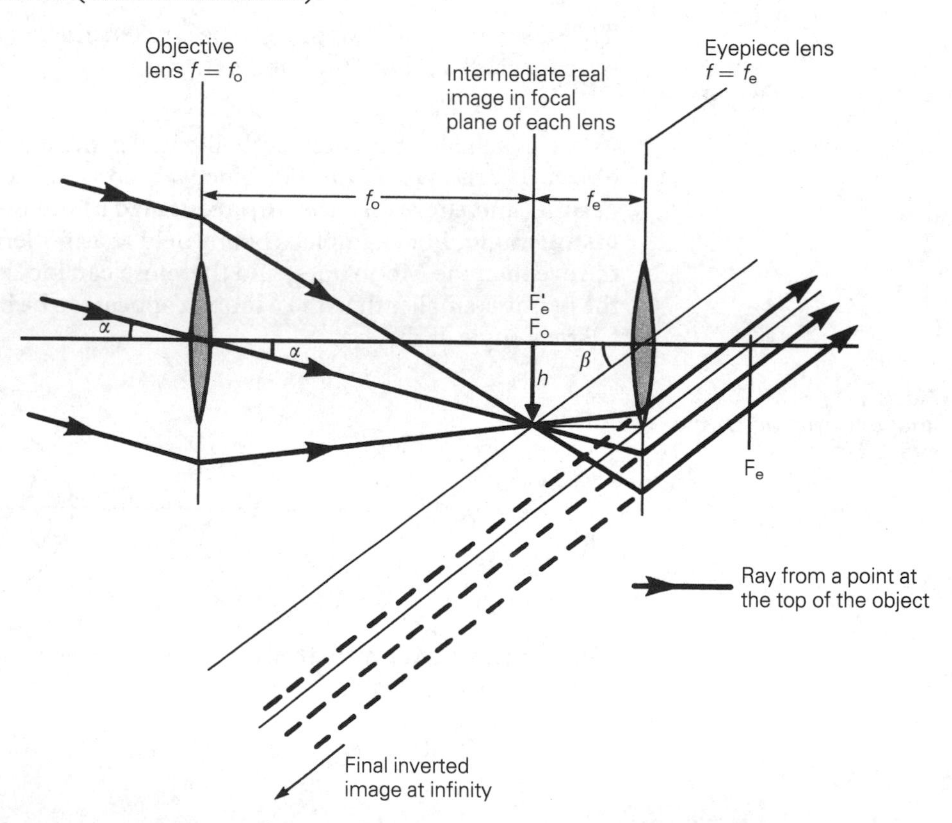

Fig. 8.2 Astronomical telescope in normal adjustment

The angular magnification, M, of a telescope is given by

$$M = \frac{\beta}{\alpha}$$

where

β = the angle subtended at the eye by the image

α = the angle subtended at the unaided eye by the object.

Since both the object and the final image are at infinity, the angles they subtend at the eye are the same as those they subtend at the objective and at the eyepiece respectively. It follows that α and β are as shown in Fig. 8.2, from which

$$\beta = \frac{h}{f_e} \quad \text{and} \quad \alpha = \frac{h}{f_o}$$

(This assumes that α and β are small, in which case, to a good approximation, $\alpha = \tan \alpha$ and $\beta = \tan \beta$.)

Therefore, since $M = \beta/\alpha$

$$M = \frac{h/f_e}{h/f_o}$$

i.e. $$M = \frac{f_o}{f_e}$$

It is clear from this equation that telescopes require long focal length objectives and short focal length eyepieces.

Final Image at the Near Point

The arrangement is shown in Fig. 8.3. The separation of the lenses is less than when the final image is formed at infinity. The intermediate image, though still in the focal plane of the objective lens, is now inside the focal point (F_e') of the eyepiece lens and in such a position that the final image is at the near point.

Fig. 8.3
Astronomical telescope with image at near point

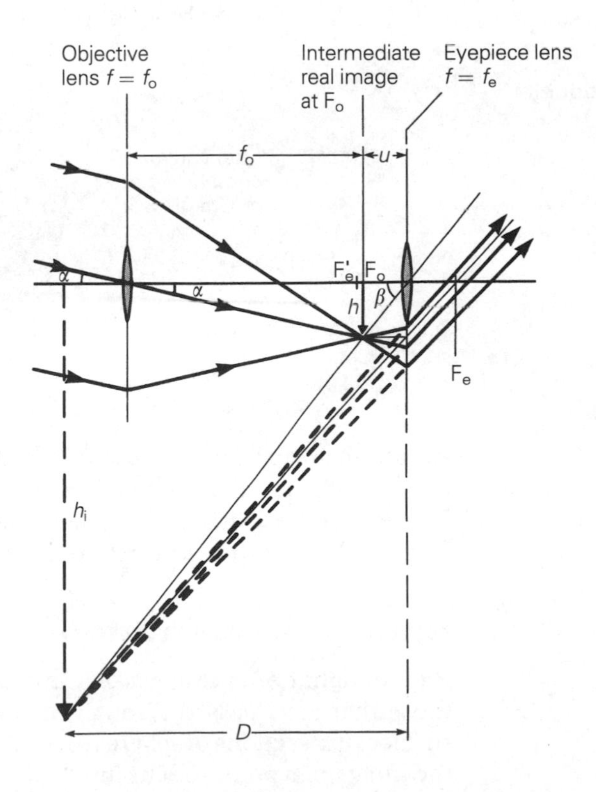

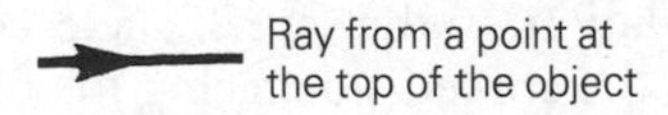

QUESTIONS 8A

1. An astronomical telescope in normal adjustment has a total length of 78 cm and produces an angular magnification of 12. What is the focal length of the objective?

2. An astronomical telescope has an objective of focal length 80.0 cm and an eyepiece of focal length 5.0 cm. The telescope is in normal adjustment and a distant object subtends an angle of 5.0×10^{-3} radians at the objective. Find: **(a)** the length of the instrument, **(b)** the angular magnification it produces, **(c)** the size of the intermediate image.

8.3 ABERRATIONS

Chromatic aberration

A simple (single-component) lens has slightly different focal lengths for the various colours that make up white light. (This is because the refractive index of glass depends on the wavelength of the light involved.) It follows that when green light (say) is in focus, the other colours are slightly out of focus and the image has a coloured fringe around it. This is called **chromatic aberration** and it can be reduced, but not totally eliminated, by using an **achromatic doublet**. This is a combination of two lenses (Fig. 8.4) each made from a different type of glass and cemented together with Canada balsam. The crown glass component is converging, and the deviation it produces is in the opposite direction to that produced by the, slightly diverging, flint glass component. Because the deviations are in opposite directions, it is possible to arrange that red light is brought to a focus at the same point as violet light.

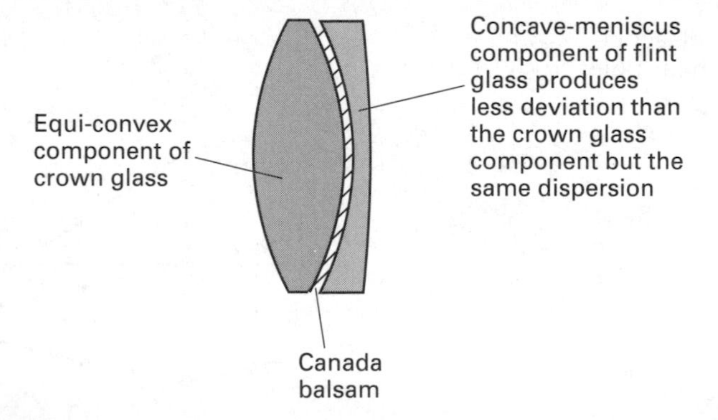

Fig. 8.4
An achromatic doublet

Although an achromatic doublet has the same focal length for red light as it does for violet light, it has slightly different focal lengths for the other spectral colours. However, these differences are considerably less than for a single-component lens and the doublet therefore produces very little chromatic aberration.

Spherical aberration

Rays of light from the edges of a lens are brought to a focus closer to the lens than those that have passed through the centre (Fig. 8.5). This occurs because lens surfaces are sections of spheres – it is known as **spherical aberration** and it causes the image of a point to blur into a disc. The image produced by a lens suffers the minimum possible spherical aberration when each surface of the lens produces the same deviation. In the case of a telescope this is achieved by using a plano-convex lens with the curved side facing the object being viewed.

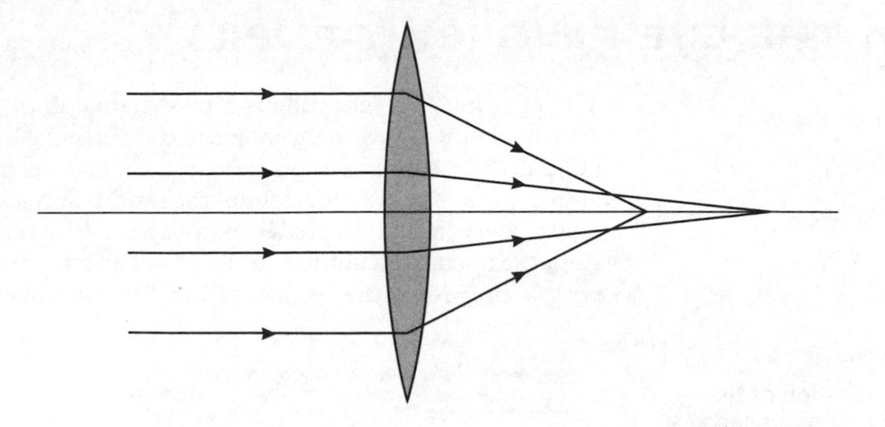

Fig. 8.5
To illustrate spherical aberration produced by a lens

The image produced by a mirror suffers spherical aberration if the surface of the mirror is spherical, but not if it is paraboloidal (i.e. of parabolic cross-section) – see Fig. 8.6.

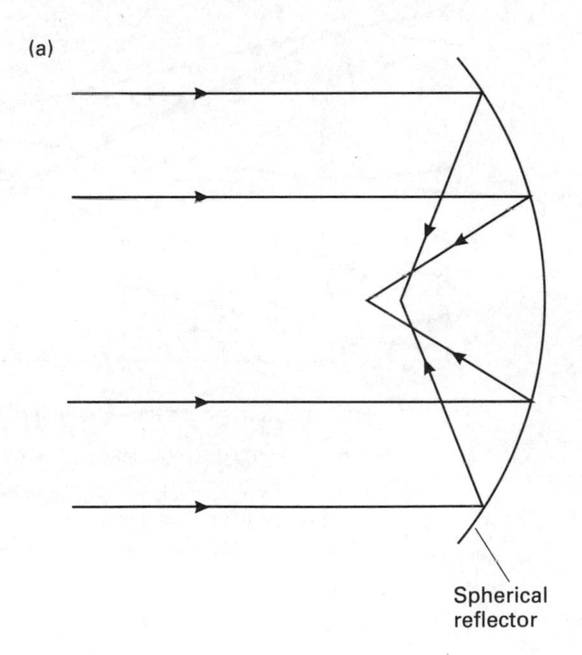

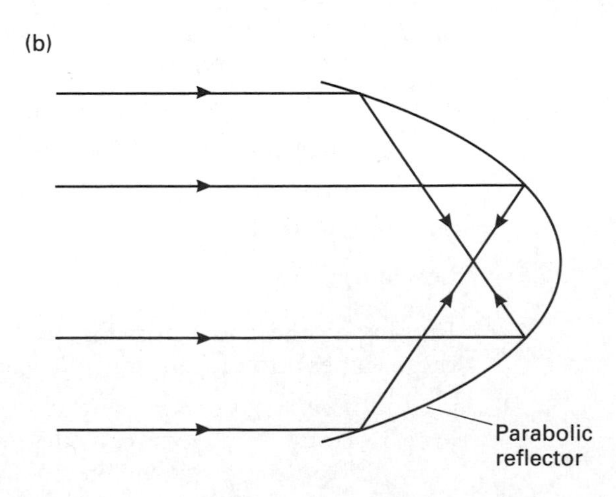

Fig. 8.6
(a) Spherical aberration produced by a spherical reflector (b) Absence of spherical aberration with a parabolic reflector (all rays brought to a common focus)

8.4 THE EYE-RING (EXIT-PUPIL)

The eye-ring of a telescope is a washer-like disc, positioned so that its circular aperture coincides with the image of the objective lens formed by the eyepiece lens (Fig. 8.7). Because of its position and because its aperture has the same diameter as this image, the eye-ring defines the **smallest** region through which passes all the light that has been refracted by both lenses. If an observer places his eye at the eye-ring, the maximum amount of light will enter his eye. **The position of the eye-ring is therefore the best position for the observer's eye.**

Fig. 8.7
The location of the eye-ring of a telescope

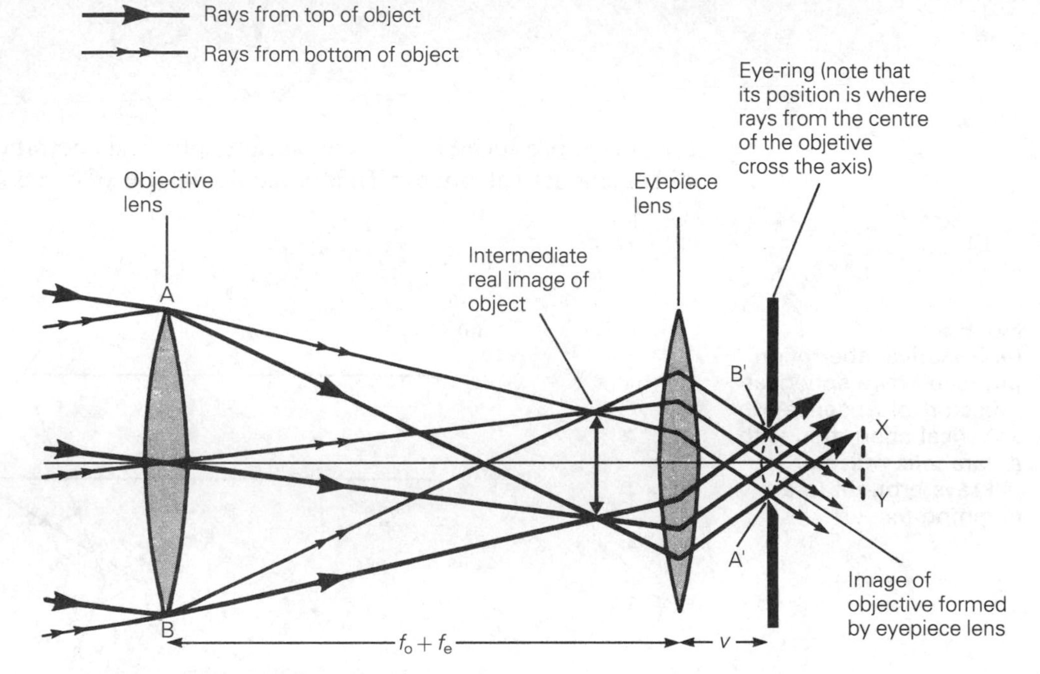

(i) Both rays from A pass through A' therefore image of A is at A'

(ii) A' B' is the smallest region through which all the light passes

(iii) XY represents the pupil of the observer's eye. Most of the light fails to enter the pupil when it is in the position shown. All the light enters the pupil when it is at A' B'

The diameter, D_e, of the eye-ring is equal to that of the image of the objective lens formed by the eyepiece lens. In normal adjustment the separation of these lenses is $f_o + f_e$ (where f_o and f_e are the respective focal lengths). Therefore, if v is the distance from the eyepiece lens to the image of the objective, and D_o is the diameter of the objective lens, then since

$$\frac{\text{Object size}}{\text{Image size}} = \frac{\text{Object distance}}{\text{Image distance}}$$

$$\frac{D_o}{D_e} = \frac{f_o + f_e}{v} \qquad [8.2]$$

Employing the usual notation, $u = f_o + f_e$, $v = v$ and $f = f_e$. Substituting these values into the lens formula leads to

$$\frac{1}{f_o + f_e} + \frac{1}{v} = \frac{1}{f_e}$$

Multiplying by $(f_o + f_e)$ gives

$$1 + \frac{f_o + f_e}{v} = \frac{f_o + f_e}{f_e}$$

Therefore, from equation [8.2]

$$1 + \frac{D_o}{D_e} = \frac{f_o + f_e}{f_e}$$

i.e. $$\frac{D_o}{D_e} = \frac{f_o}{f_e}$$

In normal adjustment the angular magnification, M, is given by $M = f_o/f_e$, and therefore

$$M = \frac{D_o}{D_e}$$

8.5 THE REFLECTING TELESCOPE

The function of the objective lens of a refracting telescope is to collect light and produce an image which can be examined by the eyepiece. In a reflecting telescope a concave mirror is used to the same end.

In the **Cassegrain** reflecting telescope (Fig. 8.8) the intermediate image (I_1) formed by the concave objective acts as a virtual object for the (small) convex mirror. This produces a second intermediate image (I_2) in a convenient position to be magnified by the eyepiece.

Fig. 8.8
Cassegrain reflecting telescope

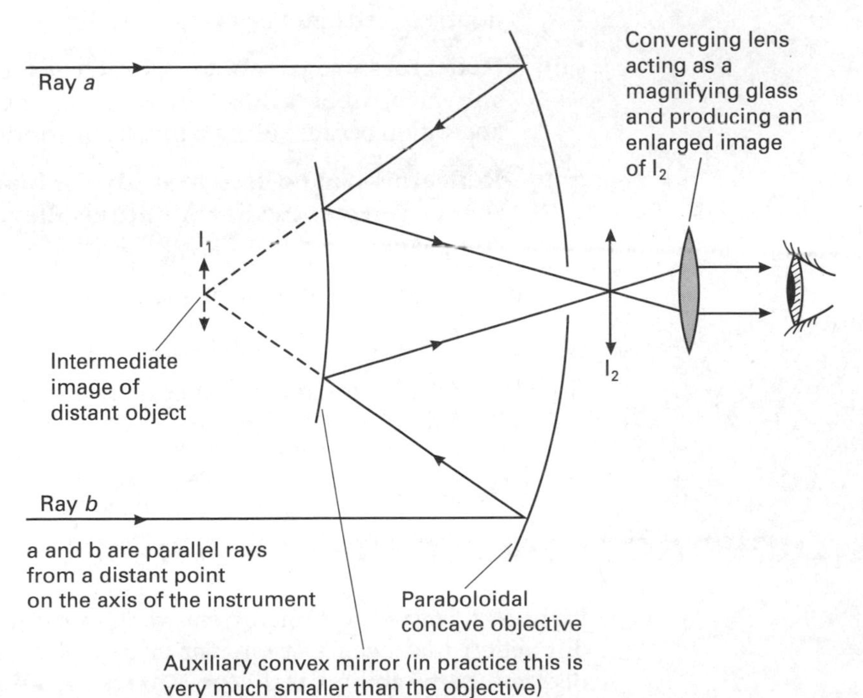

8.6 TELESCOPE OBJECTIVES

The objective lens or mirror of a telescope should have a large diameter, because:

(i) A telescope with a large-diameter objective can resolve fine detail (see section 8.9).

(ii) The amount of light gathered by a telescope is proportional to the square of the diameter of its objective.

8.7 REFRACTORS AND REFLECTORS COMPARED

Reflecting telescopes have a number of advantages over refractors.

(i) **The diameter of a mirror can be much larger than that of a lens** because a mirror can be supported over the whole of its non-reflecting surface, a lens can be supported only at its edges. As a consequence, the world's largest telescopes (i.e. those that can be used to look at very faint objects and resolve very fine detail) are reflecting telescopes. Until recently (1993) the largest was the Russian telescope on Mt. Pastukhov – its mirror has a diameter of 6 m. It has been pushed into second place by the Keck Telescope in Hawaii which has a 10 m mirror composed of 36 hexagonal segments, each of which is 1.8 m across. The world's largest refracting telescope (at the Yerkes Observatory near Chicago) has a 1 m diameter objective.

(ii) **Mirrors cannot produce chromatic aberration**. Although the chromatic aberration produced by lenses can be reduced by using achromatic doublets, it cannot be completely eliminated.

(iii) **Reflectors use paraboloidal mirrors and therefore do not produce spherical aberration**. (In practice, there is usually a little spherical aberration because of the difficulty in grinding a truly paraboloidal surface.)

(iv) **Reflectors can be used to study the long wavelength (> 300 nm) UV that penetrates the Earth's atmosphere**. Glass lenses absorb ultraviolet completely.

The main advantages of refracting telescopes are:

(i) They are less sensitive to temperature changes than reflectors.

(ii) They require less maintenance than reflectors because mirrors have to be re-aluminized periodically.

8.8 DIFFRACTION AT CIRCULAR APERTURES

Light is diffracted to some extent whenever it passes through an aperture. The diffraction that occurs at circular apertures is of particular interest because it determines the ability of telescopes to resolve fine detail.

The Fraunhofer diffraction pattern of a circular aperture is shown in Fig. 8.9. It takes the form of a bright central disc (known as **Airy's disc** and containing over 91% of the light) surrounded by a number of much less intense rings. The edge of the central disc makes an angle θ with the 'straight through' direction which is given by $\sin\theta = 1.22\lambda/D$ where D is the diameter of the aperture and λ is the wavelength of the light involved. When $D >> \lambda$, this reduces to $\theta = 1.22\lambda/D$ where θ is in radians.

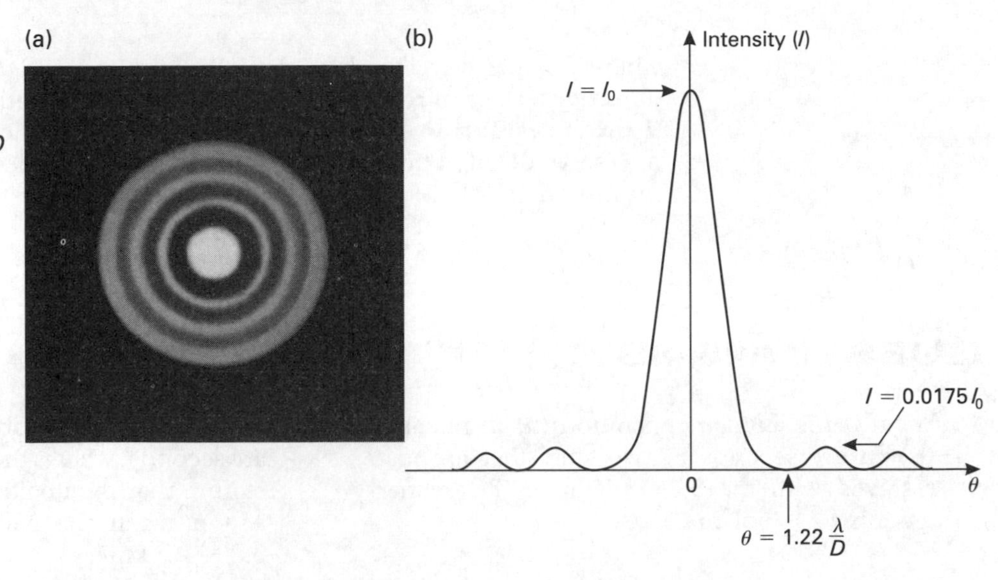

Fig. 8.9
(a) Diffraction pattern of a circular aperture
(b) Intensity distribution for aperture of diameter D

8.9 RESOLVING POWER

The resolving power of a telescope (i.e. its ability to resolve fine detail) is determined by the extent to which light is diffracted at its objective lens or mirror. The smaller the objective, the greater the diffraction and the greater the likelihood that the images of two objects will overlap and therefore be indistinguishable, i.e. unresolved.

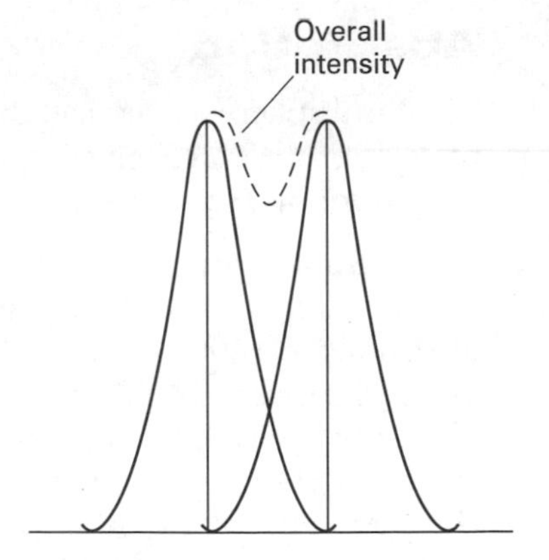

Fig. 8.10
Rayleigh's criterion

According to an arbitrary criterion proposed by Rayleigh (**Rayleigh's criterion**) two point objects will be just resolved if their angular separation is such that the central maximum in the diffraction pattern of one of them coincides with the first minimum of the other (Fig. 8.10). It follows that two point objects will be resolved by a telescope if their angular separation is not less than θ where

$$\sin\theta = 1.22\frac{\lambda}{D}$$

For small angles, i.e. when $D >> \lambda$, this reduces to

$$\theta \text{ (in radians)} = 1.22\frac{\lambda}{D} \quad (D >> \lambda) \qquad [8.3]$$

where λ is the wavelength of the light involved and D is the diameter of the objective lens or mirror. The value of θ given by equation [8.3] is known as the **limit of resolution** – the smaller the limit of resolution, the greater the ability to resolve detail. Thus, the bigger the diameter of the objective, the better the resolution.

QUESTIONS 8B

1. What is the angular separation **(a)** in radians, **(b)** in arc seconds of two stars that are just resolved with the 5.1 m Palomar Telescope? (Assume $\lambda = 550$ nm.)
2. If the limit of resolution of the human eye is 40 arc seconds, what is the diameter of the smallest feature that the unaided eye can detect on the Moon when the Moon is at a distance of 3.8×10^5 km?

Note In practice, the resolution achievable with a large telescope is likely to be worse than the theoretical (i.e. diffraction limited) value because atmospheric turbulence causes star images to jiggle around, i.e. to twinkle (see section 8.10).

8.10 ATMOSPHERIC EFFECTS

Absorption

The atmosphere absorbs all the radiation coming to us from space except in two regions of the electromagnetic spectrum known as the optical window and the radio window (section 9.1).

The optical window, which includes the whole of the **visible region** (390 nm to 780 nm), covers a band of wavelengths extending from about 300 nm in the near ultraviolet to 1400 nm in the near infrared, plus a number of narrow bands in the infrared between 1400 nm (1.4 μm) and 21 μm. The atmosphere is opaque to the rest of the infrared (21 μm to 1 mm) because of absorption, caused mainly by **water vapour** and by **carbon dioxide**. The cut-off in the ultraviolet is due to absorption by **ozone** (in the much publicized ozone layer).

X-rays and γ-rays are absorbed by gas atoms and molecules – mainly nitrogen and oxygen.

Note Some of the absorption lines seen in stellar spectra are caused by absorption in the Earth's atmosphere. These lines, unlike those due to absorption in the atmosphere of the star itself, are more pronounced when the star is low in the sky and its light has to pass through a relatively thick portion of the Earth's atmosphere.

Spectrum of the Sun showing atmospheric absorption lines when the Sun is (a) high in the sky, (b) low in the sky

Scattering

The atmosphere is not completely transparent – light is **scattered** out of the direct beam by air molecules and by dust and water droplets. **The effect is most marked when the diameter of the scattering particle is the same order of magnitude as the wavelength concerned**. Thus visible light is obscured by mist and dust to a much greater extent than is infrared radiation.

When the diameter of the scatterer is much less than the wavelength, λ, being scattered, the intensity of the scattered radiation is proportional to $(1/\lambda^4)$. This explains why the sky is blue. Air molecules scatter very much more blue light out of the direct rays of the Sun than red (because the wavelength of blue light is less than that of red light) and therefore the light that is subsequently scattered back to the ground from the rest of the sky is predominantly blue. It also explains the increasingly red colour of the Sun towards sunset. As the Sun's rays pass through greater and greater thicknesses of atmosphere, more and more blue light is scattered out of the direct beam and therefore the Sun appears more and more red.

Scattering produced by larger particles (e.g. water droplets) is independent of wavelength – an overcast sky is grey because all wavelengths are scattered to the same extent.

Refraction

The light from a star, for example, is refracted (towards the normal) as it enters the atmosphere and as it passes down through layers of ever greater density. This has the effect of making the star appear to be higher in the sky than it actually is. The closer the star is to the horizon, the greater the effect.

Refraction also accounts for the fact that stars **twinkle**. The light from a star passes through layers of air that are at different temperatures and which, therefore, have different values of refractive index, and so refract the light to different extents. As the various layers move relative to each other, the position of the star appears to jump about at random through a few arc seconds. Twinkling also involves random variations in intensity, brought about by air movements causing changes in the total mass of air in the observer's line of sight.

The increased refraction that occurs when a star is close to the horizon can make the star appear to change colour. This is because the extent to which different wavelengths are refracted changes as the layers of air move about.

Planets, as opposed to stars, twinkle only when the air is at its most turbulent. Planets appear as discs of light rather than as points and the random changes in the direction of the light coming from different parts of the disc average to zero.

8.11 TELESCOPE SITES

Ideally, an optical telescope should be sited above the clouds at a place where there is very little air turbulence and where the air is largely free from dust. Urban locations should be avoided because of the deleterious effects of street lighting (the so-called light pollution). Accordingly, telescopes are often placed on remote mountain tops. The air above mountains is often free from water vapour, making them particularly suitable sites for observations in the infrared.

Observations from Earth-orbiting satellites and from space probes are totally unaffected by the damaging effects of the atmosphere. This is costly but gives much improved images in the visible region of the spectrum, and also allows observations to be made at wavelengths that are not possible with Earth-based instruments, i.e. X-ray and γ-ray wavelengths and the bulk of the infrared and ultraviolet. Rockets and high-altitude balloons have also been used to make observations at these wavelengths.

CONSOLIDATION

The angular magnification, M, of a telescope is defined by

$$M = \frac{\beta}{\alpha}$$

where

β = the angle subtended at the eye by the image

α = the angle subtended at the unaided eye by the object.

When the final image is at infinity the telescope is in **normal adjustment** and

$$M = \frac{f_o}{f_e}$$

The objective produces a real image that acts as an object for the eyepiece. The eyepiece acts as a magnifying glass and produces a virtual image.

In deriving the expression for angular magnification it is necessary to specify that angles are small so that the approximations $\alpha = \tan\alpha$ and $\beta = \tan\beta$ may be used.

The angular separation θ of two point objects that are just resolved is given by **Rayleigh's criterion** as

$$\theta \text{ (in radians)} = 1.22\frac{\lambda}{D} \qquad (D >> \lambda)$$

Water vapour and **carbon dioxide** in the Earth's atmosphere absorb most of the **infrared** band of wavelengths. **Ozone** absorbs most of the **ultraviolet** wavelengths. **Gas atoms** and **molecules** absorb **X-rays** and **γ-rays**.

Light is **scattered** by air molecules, dust and water droplets.

Scattering is most marked when the diameter of the scattering particle is the same order of magnitude as the wavelength concerned.

Refraction by the Earth's atmosphere makes stars appear higher than they actually are. It also causes stars to twinkle and to change colour.

QUESTIONS ON CHAPTER 8

1. A telescope consists of two converging lenses: an objective lens of focal length 500 mm and an eyepiece lens of focal length 50 mm. When the telescope is in normal adjustment:
(a) what is the separation of the lenses,
(b) where is the final image located,
(c) is the image erect or inverted,
(d) what is the magnifying power,
(e) where should the pupil of the eye be placed to obtain the best view through the telescope? [S]

2. Explain the term *angular magnification* as related to an optical instrument.
Describe, with the aid of a ray diagram, the structure and action of an astronomical telescope. Derive an expression for its angular magnification when used so that the final image is at infinity. With such an instrument what is the best position for the observer's eye? Why is this the best position?
Even if the lenses in such an instrument are perfect it may not be possible to produce clear separate images of two points which are close together. Explain why this is so. Keeping the focal lengths of the lenses the same, what could be changed in order to make the separation of the images more possible? [L]

3. Define the *angular magnification* (or *magnifying power*) of a telescope. A telescope consists of two converging lenses. When in normal adjustment, so that the image of a distant object is formed at infinity, the lenses are 450 mm apart, and the angular magnification is 8. What are the focal lengths of the two lenses? Is the image erect or inverted? [S]

4. **(a)** What is meant by *normal adjustment* for an astronomical telescope? Why is it used in this way?
(b) An astronomical telescope in normal adjustment is required to have an angular magnification of 15. An objective lens of focal length 900 mm is available.
Calculate the focal length of the eyepiece required and draw a ray diagram, not to scale, to show how the lenses should be arranged. The diagram should show three rays passing through the telescope from a non-axial point on a distance object. State the position of the final image, and whether or not it is inverted. [N]

5. Define *magnifying power* of an optical telescope. Draw a ray diagram for an astronomical refracting telescope in normal adjustment, showing the paths through the instrument of three rays from a non-axial distant point object. Derive an expression for the magnifying power. The magnifying power of an astronomical telescope in normal adjustment is 10. The real image of the *objective lens* produced by the eye lens has a diameter 0.40 cm. What is the diameter of the objective lens?
Discuss briefly the significance of the diameter of the objective lens on the optical performance of a telescope. [N]

6. State what is meant by the *normal adjustment* in the case of an astronomical telescope.
Trace the paths of three rays from a distant non-axial point source through an astronomical telescope in normal adjustment.
Define the *magnifying power* of the instrument, and by reference to your diagram, derive an expression for its magnitude.
A telescope consists of two thin converging lenses of focal lengths 100 cm and 10 cm respectively. It is used to view an object 2000 cm from the objective. What is the separation of the lenses if the final image is 25 cm from the eye-lens? Determine the magnifying power for an observer whose eye is close to the eye-lens. [N]

7. Draw a ray diagram showing the action of an astronomical telescope, consisting of two thin converging lenses, in normal adjustment, when forming separate images of two stars.

Explain why the images of the two stars formed on the retina of the eye will have a greater separation if the stars are viewed through the telescope than if they are viewed by the unaided eye.
The objective of an astronomical telescope in normal adjustment has a diameter of 150 mm and a focal length of 4.00 m. The eyepiece has a focal length of 25.0 mm. Calculate:
(a) the magnifying power of the telescope,
(b) the position of the eye-ring (that is, the position of the image of the objective formed by the eyepiece),
(c) the diameter of the eye-ring.
Give one advantage of placing the eye at the eye-ring. [L]

8. **(a)** **(i)** What is meant by the magnifying power M of an astronomical telescope?
(ii) Such a telescope can be made from two converging lenses. With the aid of a sketch, derive an expression for M in terms of the focal length of these lenses. Assume that the telescope is in normal adjustment.
(b) **(i)** Explain the significance of the eye-ring of a telescope.
(ii) The diameters of eye-rings seldom exceed 4 mm. Suggest a reason why.
(c) A telescope objective lens (of focal length 2.00 m) is used to photograph the Moon, which subtends an angle of 9.2 mrad (= 0.53°) at the Earth. The photographic film is placed in the principal focal plane of the lens, at right-angles to its principal axis.
(i) Calculate the diameter of the image of the Moon formed on the photographic film.
(ii) Suggest a reason why the complete telescope in normal adjustment is not used for taking the photograph.
[O*, '92]

9. Draw a ray diagram to show how a converging lens produces an image of finite size of the Moon clearly focused on a screen. If the Moon subtends an angle of 9.1×10^{-3} radian at the centre of the lens, which has a focal length of 20 cm, calculate the diameter of this image.
With the screen removed, a second converging lens of focal length 5.0 cm is placed coaxial with the first and 24 cm from it on the side remote from the Moon. Find the position, nature and size of the final image. [N]

10. Define *magnifying power* of a telescope. Show that, if a telescope is used to view a distant object and is adjusted so that the final image is at infinity, the magnifying power is given by the ratio of the focal lengths of the objective and eyepiece.
An astronomer has the choice of two telescopes of equal magnifying power but of different apertures. Explain what advantages he could obtain by choosing one rather than the other.
A telescope has two lenses of focal lengths 1.0 m and 0.10 m and it is adjusted to produce an image of a distant object on a screen. The object subtends an angle of 0.30° at the telescope objective. Calculate: **(a)** the linear size of the image formed on the screen 0.5 m from the eyepiece, and **(b)** the distance between the two lenses. Draw a diagram of the optical arrangement showing the paths of *two* rays through the lens system which come from a point on the object not on the axis.
[L]

11. **(a)** Draw a ray diagram to show how a thin converging lens can form a magnified virtual image of an object.
What is meant by the term *angular magnification* and what is its value in this case?
Explain why the image appears magnified when both object and image subtend the same angle at the lens.
(b) A telescope is made from two lenses of focal lengths 100 cm and 5 cm. The instrument is adjusted so that a virtual image of the Moon is formed 25 cm from the more powerful lens.
(i) Draw a ray diagram of this arrangement.
(ii) Calculate the distance between the lenses and the angular magnification produced by the instrument.
(iii) Explain where the observer should position his eye in order to get the greatest field of view when using the telescope.
(c) The above telescope is now adjusted to form a real image of the Moon 20 cm from the more powerful lens.
(i) Calculate the distance between the lenses.
(ii) Suggest a possible use for a telescope adjusted in this manner. [S]

12. **(a)** **(i)** Draw a ray diagram of a Cassegrain reflecting telescope, showing the paths of two parallel rays from a distant object on the axis of the telescope to the eyepiece lens.
(ii) Giving a reason, state **one** optical advantage (apart from any involving aberrations) which a reflecting telescope has over a refracting telescope.
(b) **(i)** With the aid of a ray diagram show how *spherical aberration* arises in the case of a spherical mirror.
(ii) State how spherical aberration may be eliminated in a reflecting telescope. [N, '96]

13. A distant object subtending an angle of 0.10 minute of arc is viewed with a reflecting telescope whose objective is a concave mirror of focal length 1000 cm. The reflected light is intercepted by a convex mirror placed 950 cm from the pole of the concave mirror and a real image is formed in the vicinity of the pole of the concave mirror where there is a hole. This imaged is viewed with a convex lens of focal length 5 cm used as a magnifying glass and producing a final image at infinity. Draw a ray diagram (not to scale) for this arrangement using two rays from a non-axial point on the distant object which strike the objective at a small angle with the principal axis of the system.
Calculate:
(a) the size of the real image that would have been formed at the focus of the concave mirror,
(b) the size of the image formed by the convex mirror,
(c) the angle subtended by the final image at the optic centre of the convex lens.
Give **two** advantages of reflecting telescopes over refracting telescopes. (1 minute of arc $= 2.9 \times 10^{-4}$ radians.) [N]

14. The absorption spectrum of the Sun was recorded with a high resolution spectrometer on a fine cloudless day. Two spectra were recorded, one at about noon and the second a few hours later. Explain why the spectrum recorded at the later time showed more dark lines than the spectrum recorded at the earlier time. [N*]

15. Suggest how an optical reflecting telescope with a main mirror of diameter about 2 m can be adapted to detect infrared radiation. Explain why it is necessary for infrared telescopes to be either sited at a high altitude or flown in balloons. [N*]

16. **(a)** Stars emit radiation at all wavelengths, but most regions of the electromagnetic spectrum cannot be observed from the Earth's surface. Explain why this is so for γ-rays, X-rays, ultraviolet radiation and infrared radiation.
State how observations at these wavelengths can be made.
(b) Radiation in the visible region of the electromagnetic spectrum is scattered in the atmosphere. State what causes this scattering and under what conditions the scattering will be most significant. [N*, '93]

17. Telescopes, such as the Hubble telescope, which orbit the Earth have enabled scientists to photograph astronomical objects. These photographs can provide far more information than photographs obtained using Earth-based telescopes. State reasons for this improvement. [C*, '96]

9

RADIO AND RADAR ASTRONOMY

9.1 RADIO ASTRONOMY

Radio astronomy allows us to study certain extraterrestrial objects by observing the electromagnetic radiation they emit at radio frequencies. Observations, unlike those at visible wavelengths, are not limited to night-time and are not affected by cloud cover.

The so-called **radio window** extends from about 1 cm to 15 m. Wavelengths below 1 cm are absorbed by water vapour in the atmosphere; those above 15 m are reflected by the ionosphere. In order to reduce the problems caused by interference from radar and microwave communication systems, certain frequencies have been set aside specifically for radio astronomy. The radio window is very much wider than the optical window – the upper wavelength limit is 1500 times the lower limit, the corresponding value for the optical window is less than 5.

Sources of radio emission include the Sun, pulsars, quasars and supernova remnants. The 21 cm radio emission (see 'Note' below) from atomic hydrogen has proved an invaluable means of determining the structure of the Galaxy. Vast clouds of dust in the galactic plane limit our view at optical wavelengths because visible light is **scattered** (section 8.10) by the dust. Radio waves can penetrate the dust (because they have very much longer wavelengths) and therefore can provide information about the core of the Galaxy.

Note Particles such as protons and electrons have a property called **spin**. Spin is a quantum mechanical concept – it has no classical analogue but it is sometimes helpful to think of a particle as a tiny sphere spinning about an axis through its centre. If the electron and proton in a hydrogen atom are spinning in the same direction, the electron has a little more energy than when they are spinning in opposite directions. Collisions between atoms maintain an equilibrium in which more atoms are in the low-energy state than the high-energy state. If an electron in the high-energy state drops into the lower state, a radio-frequency photon with a wavelength of 21 cm is emitted.

9.2 SINGLE DISH RADIO TELESCOPES

A radio telescope (Fig. 9.1) consists of a large (up to 100 m) parabolic dish which collects the incoming radio waves and reflects them on to a dipole antenna at its focus. The electrical signal induced in the antenna is fed along cables to amplifiers and recording equipment in the control room, usually at the base of the telescope. The signals detected by a radio telescope are often very weak and therefore low-noise, high-gain amplifiers are required.

Fig. 9.1
Radio telescope

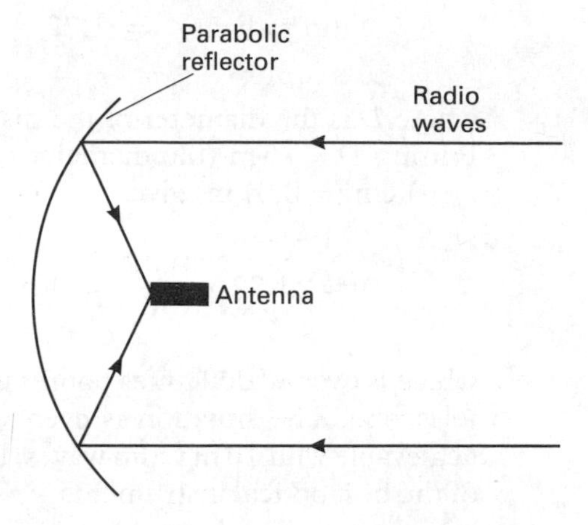

Only those radio waves that are contained in a narrow beam along the axis of the telescope are focused on to the antenna. In order to build up a radio 'picture', the telescope has to be scanned to and fro, and up and down across the radio source that is being observed.

The reflecting surface of the dish is either sheet metal or wire mesh, mounted in such a way that it maintains its shape when pointed at different regions of the sky. Wire mesh, which is lighter, can be used when the wavelength of the radio waves being studied is greater than about 20 times the mesh size. If the wavelength is less than this, the image is degraded by diffraction effects.

The Lovell Radio Telescope at Jodrell Bank, Cheshire. The dish has a diameter of 76 metres.

The dish should have a large diameter, for basically the same reasons that an optical telescope should have a large-diameter objective, i.e. because:

(i) The rate at which the dish collects (radio) energy is proportional to its cross-sectional area and therefore to the square of its diameter.

(ii) The bigger the diameter, the greater the ability to resolve detail.

The ability of a radio telescope to resolve detail is very much poorer than that of an optical telescope. The smallest angular separation, θ, of two points that can be resolved by a radio telescope (the **limit of resolution**) is given by analogy with equation [8.3] as

$$\theta \text{ (in radians)} = 1.22\frac{\lambda}{D}$$

where D is the diameter of the dish and λ is the wavelength of the radio waves. Putting $D = 76\,\text{m}$ (the diameter of the dish of the Jodrell Bank telescope) and $\lambda = 1\,\text{cm}$ ($= 0.01\,\text{m}$) gives

$$\theta = 1.22 \times \frac{0.01}{76} = 1.6 \times 10^{-4} \text{ radians}$$

which is over a 1000 times poorer than the corresponding value for a large optical telescope. The situation is even worse at longer wavelengths – the resolution achievable with 10 m radio waves, for example, is a million times poorer than that of the best optical instruments.

EXAMPLE 9.1

A quasar is 6.5×10^{25} m from the Earth and emits radio-frequency energy at a rate of 2.0×10^{40} W. Calculate the power of the signal received by a radio telescope that has a dish with a diameter of 64 m, assuming that the quasar radiates uniformly in all directions.

Solution

Area of a sphere $= 4\pi r^2$ Area of a circle $= \pi r^2$

The energy radiated by the quasar is spread over the surface of a sphere of radius 6.5×10^{25} m, i.e. over an area of $4\pi\,(6.5 \times 10^{25})^2\,\text{m}^2$.

The dish has a radius of 32 m and therefore collects this energy over an area of $\pi(32)^2\,\text{m}^2$.

It follows that

$$\frac{\text{Power received}}{\text{Total power emitted}} = \frac{\pi(32)^2}{4\pi(6.5 \times 10^{25})^2}$$

$$\therefore \quad \frac{\text{Power received}}{2.0 \times 10^{40}} = \frac{\pi(32)^2}{4\pi(6.5 \times 10^{25})^2}$$

$$\text{i.e.} \quad \text{Power received} = \frac{2.0 \times 10^{40} \times (32)^2}{4 \times (6.5 \times 10^{25})^2} = 1.2 \times 10^{-9}\,\text{W}$$

9.3 OPTICAL REFLECTING TELESCOPES COMPARED WITH RADIO TELESCOPES

Similarities

(i) The function of both radio telescopes and optical telescopes is to detect electromagnetic radiation.

(ii) Both use a parabolic concave reflector to collect the radiation and to focus it on to a detector.

(iii) The rate at which each type collects radiation is proportional to the square of the diameter of its reflector.

(iv) Both have to have a mechanism that allows them to track the object under observation in order to counter the effect of the Earth's rotation.

Differences

(i) Radio telescopes have to scan the object under observation in order to form an image of it.

(ii) Radio telescopes are used to detect very much longer wavelengths than optical telescopes and therefore can 'see' through clouds of interstellar dust.

(iii) The resolving power of a radio telescope is very much poorer than that of an optical telescope.

(iv) The resolving power of a radio telescope is limited by diffraction effects, that of a large Earth-based optical telescope is limited by the effects of turbulence in the Earth's atmosphere.

9.4 RADAR ASTRONOMY

Radio astronomy involves the detection of radio waves that have been emitted naturally by various sources in the Universe. **Radar astronomy**, on the other hand, is concerned with transmitting microwave radio signals from Earth and detecting the signals reflected back to the Earth by various objects in the Solar System. It cannot be used for objects much beyond the orbit of Saturn, because the reflected beam is so spread out that it is too weak to detect. Wavelengths are typically in the range 1 to 20 cm.

Radar astronomy provides a very accurate way of determining the distances to the Moon and the inner planets. Suppose the interval between transmitting and receiving a radar pulse reflected from the surface of Venus is 800 s. The radar beam, travelling at the speed of light (3.00×10^8 m s^{-1}), has taken half this time (i.e. 400 s) to reach Venus and therefore the distance from the Earth to Venus at the time of the measurement is $3.00 \times 10^8 \times 400 = 1.20 \times 10^{11}$ m.

Once the distance of a planet is known, its diameter can be found by measuring the angle it subtends at the Earth – see Fig. 9.2.

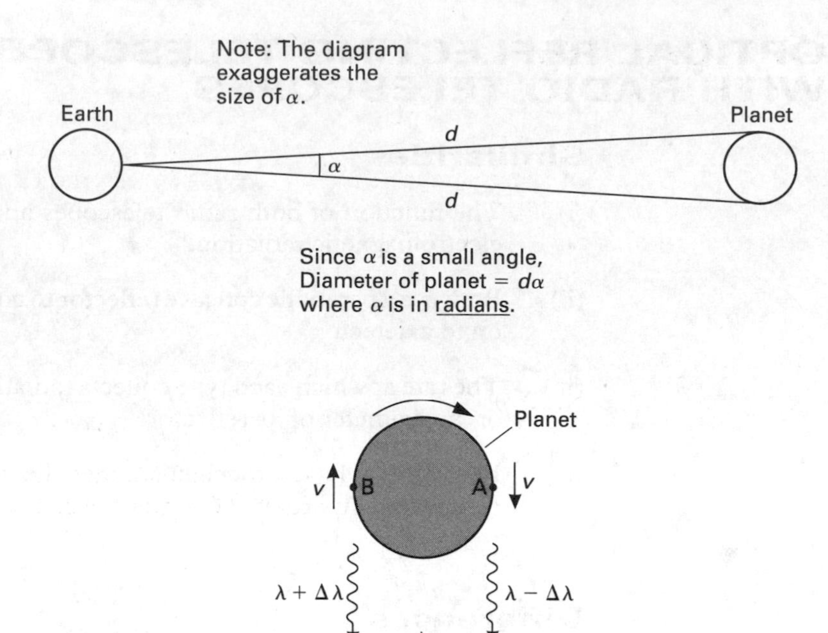

Fig. 9.2
Determining the diameter of a planet

Fig. 9.3
Using the Doppler effect to determine the speed of rotation of a planet

The speed at which a planet rotates about its axis can be found from the extent to which reflected radar waves are Doppler broadened. Suppose a radar beam with a precise wavelength, λ, is reflected from a planet as shown in Fig. 9.3. The waves that return to the Earth from the side of the planet that is approaching the Earth are shifted to shorter wavelengths; those from the other side are shifted to longer wavelengths. The signal received at the Earth therefore covers a band of wavelengths centred on λ. The largest shift in wavelength occurs for waves reflected from the edges of the planetary disc, i.e. from A and B. If A is approaching the Earth with velocity v, the waves received from A are shifted by $\Delta\lambda$, where

$$\frac{\Delta\lambda}{\lambda} = \frac{2v}{c} \qquad (v << c)$$

which enables v to be found. (**Note**: The shift is twice that given by equation [6.1] because the beam has been reflected.) If the planet's radius, r, is known, its period, T, can be found from

$$T = \frac{2\pi r}{v}$$

Notes

(i) Venus is covered by thick clouds that visible light cannot penetrate. Because it has a much longer wavelength, radar penetrates the cloud and has allowed us to map the surface of the planet and to determine the speed at which it spins about its axis.

(ii) It used to be thought that Mercury always keeps the same side facing the Sun, i.e. that its period of rotation about its own axis is the same as that of its rotation about the Sun. Radar measurements made in 1965 proved that this is not so.

QUESTIONS 9A

1. A radar beam with a wavelength of 20.000 000 cm is aimed at Mars. The wavelengths of the reflected beam range from 19.999 968 cm to 20.000 032 cm. Calculate the period of rotation of Mars. (Radius of Mars = 3.4×10^6 m, speed of light = 3.0×10^8 m s^{-1}.)

2. Explain why it is necessary to use a pulsed beam when using radar to determine the distance between the Earth and a planet.

CONSOLIDATION

The radio window extends from about 1 cm to 15 m. Wavelengths below 1 cm are absorbed by water vapour in the atmosphere; those above 15 m are reflected by the ionosphere.

Radio waves, unlike visible light, are not scattered by interstellar dust (because their wavelengths are very much larger than the diameters of the dust particles) and therefore radio astronomy allows us to study sources that cannot be seen at visible wavelengths.

Radio astronomy involves the detection of radio waves that have been emitted naturally by various sources in the Universe.

Radar astronomy involves transmitting radio signals from Earth and detecting the signals reflected back to the Earth by various objects in the Solar System.

QUESTIONS ON CHAPTER 9

1. **(a)** Explain why radio telescopes and reflecting optical telescopes have essentially similar designs. Discuss briefly **two** similarities and **two** differences between these types of telescope.
   (b) State what factors determine the resolving power of a telescope and hence estimate the resolving power of an optical telescope having a mirror of diameter 4.0 m. Show by calculation that it is impossible in practice to have a single radio telescope of the same resolving power. [N*, '92]

2. **(a)** The dish of a radio telescope has holes of diameter 0.02 m drilled at close intervals in its reflecting surface in order to reduce the weight of the dish. Explain why the performance of this radio telescope will be far less satisfactory when receiving signals of frequency 1.5×10^{10} Hz than when receiving signals of frequency 7.5×10^8 Hz.
   (b) A radio telescope with a 50 m diameter dish receives a radio signal of power 4.0×10^{-13} W from a radio source at a distance of 700 kiloparsecs.
   (i) Calculate the total power emitted by the radio source within the range of frequencies which the telescope receives, assuming that the radio source radiates uniformly in all directions.
   (ii) Calculate the distance of the radio source from the Earth in light-years and state, with reasons, whether this source lies within our own Galaxy.
   (Speed of light = 3.0×10^8 m s^{-1}, 1 kiloparsec = 3.0×10^{19} m.)
   (c) Radio signals have been detected from the Galactic centre although no details of the stars near the Galactic centre have ever been recorded at optical wavelengths. What information can be deduced from these observations about the contents of the region surrounding the point where the radio signals originated? [N]

3. **(a)** A radio telescope, when directed in the plane of the Milky Way, detects a continuous signal. The precise frequency detected varies slightly with direction in this plane and the intensity also changes.
(i) Calculate the wavelength of the radiation corresponding to a frequency of 1428 MHz.
(ii) State the most likely physical origin of this radiation.
(iii) State to what use the detection of this radiation has been put, and describe any results obtained.
(b) Explain briefly how radio astronomy and radar astronomy differ. Describe how radar astronomy can be used to determine the distance of a planet from Earth. (Speed of light = 3.00×10^8 m s^{-1}.) [N*]

4. **(a)** Discuss the similarities and differences between a steerable dish radio telescope and an optical reflecting telescope.
(b) What factors determine the largest and smallest wavelengths at which a ground-based radio telescope can be used?
(c) A weak radio signal at a frequency of 1420 MHz is observed by a radio telescope when it is directed in the Galactic plane.
(i) Calculate the wavelength of this radiation.
(ii) Suggest the most likely cause of this radiation.
(Speed of light = 3.00×10^8 m s^{-1}.) [N]

5. Describe how *radar astronomy* is used to measure the distance between a planet and Earth.
(a) Explain why the wavelengths used in radar are confined to the range 1 cm to 20 cm.
(b) What information concerning the planets, other than distance, can be obtained using radar astronomy?
(c) Explain why radar cannot be used to take measurements of a star such as α Centauri which is a distance of 1.3 parsec from Earth. [N*, '91]

6. **(a)** Explain how observation of reflected radar pulses from the surface of Venus can be used to measure
(i) the distance of Venus from the Earth,
(ii) the velocity of a point on the surface of Venus at its equator.
(b) Explain why the surface features of Venus cannot be mapped using an optical telescope but can be mapped using a radar method.
(c) A radio telescope with a dish aerial is modified to transmit and receive radar pulses of short duration. When the aerial is pointed directly at Venus, the time elapsing between the transmission of a radar pulse and the reception of the reflected pulse from Venus has a minimum value of 280 seconds. If the experiment is repeated later, the time interval between transmission of a pulse and reception of the reflected pulse reaches a maximum value of 1720 seconds. After a further period the time interval is again found to be at its minimum value.
Give expressions for the minimum and maximum distances between the Earth and Venus in terms of their orbital radii, and hence determine the radius of the orbits of Venus and the Earth, assuming that both orbits are in the same plane, concentric and circular.
The speed of electromagnetic waves in vacuo = 3.00×10^8 m s^{-1}. [N]

7. Radio waves of certain frequencies emitted from galaxies cannot be observed on Earth. Explain why this is so. [N*, '93]

8. A radar beam with a wavelength of 20.0 cm is sent towards Venus. The signal reflected back to the Earth has a total spread in wavelength of 4.82×10^{-7} cm. Calculate the period of rotation of Venus.
(Radius of Venus = 6.05×10^3 km, speed of light = 3.00×10^8 m s^{-1}.)

10

OPTICAL DETECTORS

10.1 THE EYE AND ITS FUNCTION

An eye (Fig. 10.1) produces a real, inverted image of the object being viewed. The image is produced on the **retina** – the light-sensitive region at the back of the eye. The shape, and therefore the focal length, of the eye **lens** can be altered by the action of the **ciliary muscles** attached to it. This makes it possible for light from objects which are at different distances from the eye to be brought to a focus on the retina even though it is at a fixed distance from the lens. This ability of the eye is known as **accommodation**. (Compare this with the action of a camera, where objects at different distances are focused by altering the distance between the lens and the film.)

Fig. 10.1
Section through an eye

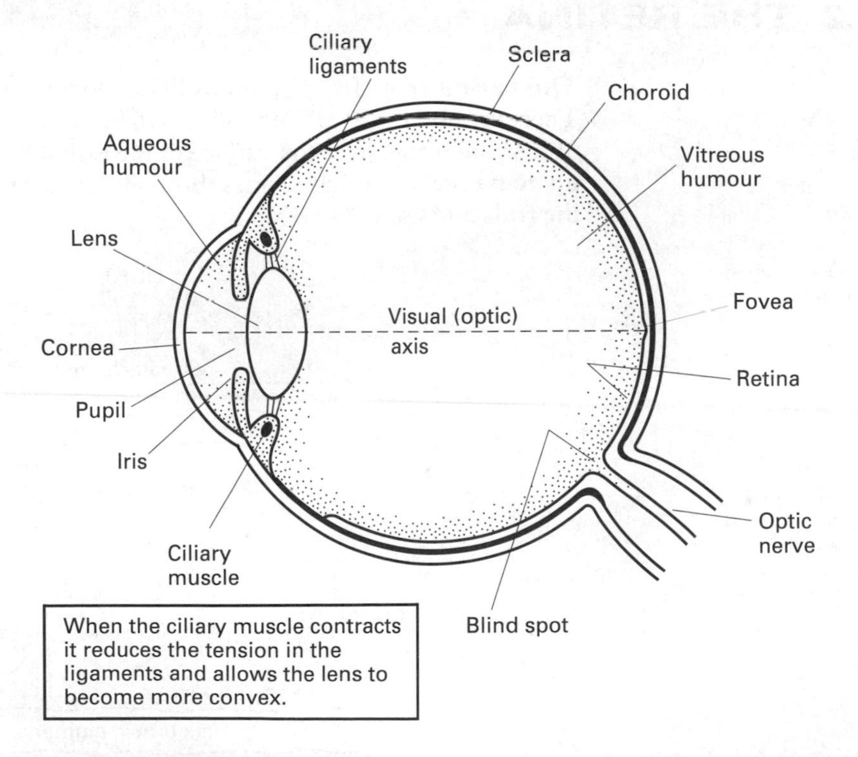

The aqueous humour (a watery, transparent fluid) and the **vitreous humour** (a jelly-like, transparent fluid) provide nutrients and help to maintain the shape of the eyeball. The refractive index of the lens (Table 10.1) is not very different from that of the aqueous humour or that of the vitreous humour. Consequently light undergoes very little deviation as a result of passing through the lens itself. **The main function of the lens is to provide slight changes in deviation rather than large amounts of it. Most of the deviation occurs at the boundary**

between the air and the cornea – where there is the largest change in refractive index.

Table 10.1
The refractive indices of the optical components of the eye

Medium	*Refractive index*
Air	1.00
Cornea	1.38
Aqueous humour	1.34
Surface of lens	1.39
Centre of lens	1.41
Vitreous humour	1.34

The iris is a (pigmented) diaphragm that controls the intensity of the light reaching the retina. It does this by adjusting the size of the **pupil** in response to signals from the retina – a negative feedback mechanism.

The sclera (the white of the eye) is the eye's tough outer cover. Its inner lining, the **choroid,** provides the blood supply for the retina. It contains a large amount of black pigment in order to reduce reflection of light within the eye and so prevent blurring of the image. The point at which the **optic nerve** joins the eye (from the brain) is known as the **blind spot** – so called because there are no light-sensitive receptors there.

10.2 THE RETINA

The retina (Fig. 10.2) contains light-sensitive cells known as **rods** and **cones.** There are about 120 million rods and 6 million cones. They are connected to nerve fibres which pass over the surface of the retina before joining together to form the **optic nerve**. Light has to pass through this layer of nerves before it can stimulate the rods and cones!

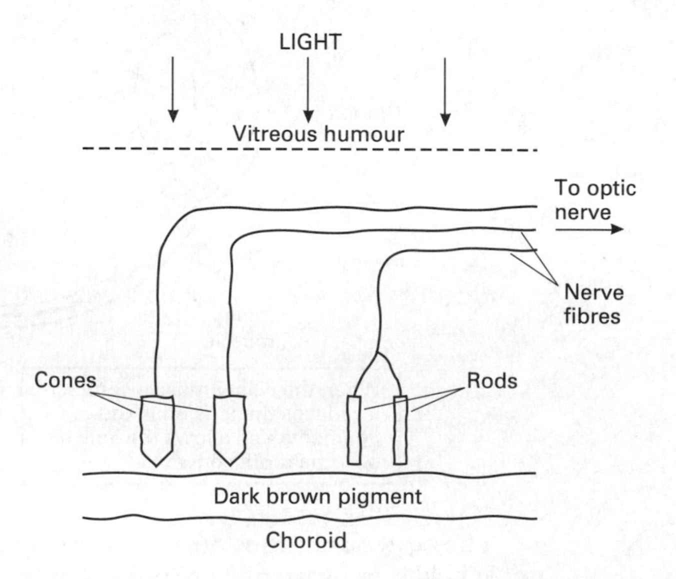

Fig. 10.2
The retina

Rods cannot distinguish different colours but they allow us to see at much lower light intensities than cones because they have greater sensitivity to light and because several of them are connected to a single nerve fibre. (The fibre is triggered by the cumulative effect of stimuli which individually would be too weak.) Because they are grouped in this way, they give little perception of detail.

Cones allow us to distinguish different colours because there are three types of them, each sensitive to a different colour. They share fewer nerve fibres than rods and therefore allow us to see fine detail. **Cones are effective only in bright light – which is why we cannot see colours when the light is dim**. This explains why two stars which are clearly of different colours when seen through a telescope appear to be the same colour when seen with the naked eye.

Rods and cones are not evenly distributed over the retina. The **fovea** (Fig. 10.1) contains only cones, but the proportion of rods increases with distance from the fovea so that the edge of the retina consists almost entirely of rods. It follows that **rods are responsible for our peripheral vision**. A faint star, for example, is more likely to be seen by looking slightly to the side of it so that the image is formed on the rods on the periphery of the retina, rather than on the cone-rich fovea. The cones in the fovea are each connected to a single nerve fibre. Because of this and because there are no nerve fibres covering it (they are arranged radially around it), **the fovea is the region of most acute vision**.

10.3 ADAPTATION OF THE RETINA

The pupil can vary in diameter from about 1.5 mm to about 8 mm, corresponding to a change in area, and therefore in the amount of light entering the eye, of a factor of about 30. Change in pupil size is primarily a response to sudden changes in illumination, it cannot account for the fact that we can see relatively well in moonlight and in sunlight, which can be over 10^5 times brighter. The ability of our eyes to adapt in this way is due mainly to a variation in the sensitivity of the retina itself. There are two processes – dark adaptation and light adaptation.

Dark adaptation

Dark adaptation is the process by which the sensitivity of the retina increases when the light intensity decreases.

Rods and cones contain pigments which decompose on exposure to light, stimulating the associated nerve cells in the process. Enzyme activity subsequently regenerates the pigments – slowly in the rods, more quickly in the cones. In very bright light the pigment in the rods is broken down so much more quickly than it can be regenerated that only the cones remain active. If the level of illumination suddenly falls, vision is difficult at first because the rods are inactive. (It is particularly difficult if the intensity is so low that the cones are also inactive!) However, because the rate of decomposition is now much lower, the pigment concentration gradually builds up and so increases the sensitivity of the retina. Dark adaptation is more or less complete after about 30 minutes.

Light adaptation

Light adaptation is the process by which the sensitivity of the retina decreases when the light intensity increases.

Sudden exposure to a higher level of illumination causes the photosensitive pigments in the rods and cones to decompose at an increased rate. Since regeneration is a slow process (particularly in rods), it cannot keep up with the rate of decomposition and therefore the number of active rods and cones falls and the sensitivity of the retina decreases. It is a very much quicker process than dark adaptation – the eyes take only a few minutes to adapt to bright sunlight, for example.

10.4 SPECTRAL RESPONSE OF THE EYE

The eye is able to distinguish different colours because it has three types of cone, each containing a different pigment and therefore each responsive to a different colour – one primarily to red, one to green and one to blue (Fig. 10.3). When light of any particular wavelength falls on the retina, the extent to which it stimulates the various cones determines the colour perceived by the brain. (**Note** that the overall sensitivity of the cones has its maximum value at about 555 nm.)

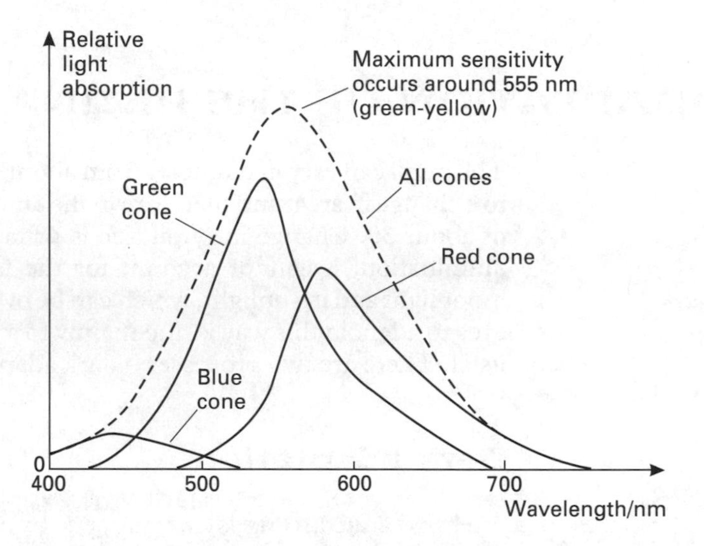

Fig. 10.3
Relative sensitivity of the three types of cone

Although the rods provide no information about colour, they are not equally sensitive to all wavelengths – they have a peak response in the blue-green region around 510 nm. The spectral response of the rods is compared with that of the cones in Fig. 10.4.

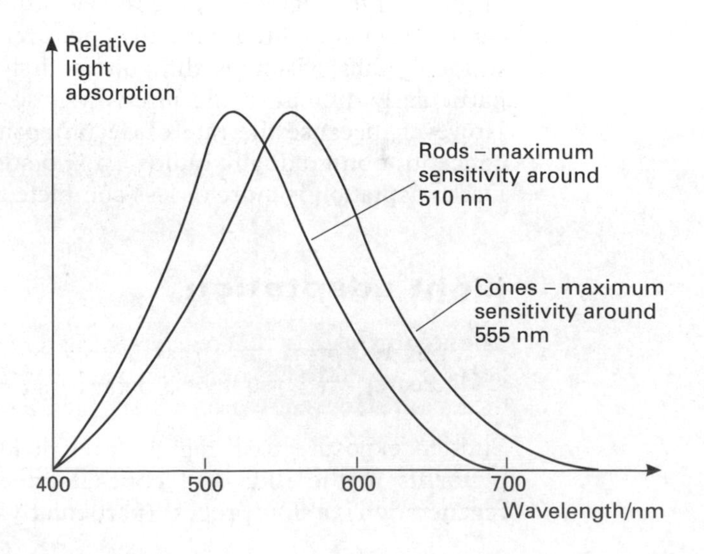

Fig. 10.4
Spectral absorption of rods and cones

QUESTIONS 10A

1. The wavelength at which the eye has its peak response depends on the intensity of the illumination. State whether a decrease in light intensity shifts the peak response to longer or to shorter wavelengths. Explain your answer.

10.5 THE CAMERA

A simple camera is shown in Fig. 10.5. The lens can be moved in and out relative to the film so that light from objects at different distances may be focused on the film. The diameter, *d*, of the aperture can be altered by means of the diaphragm adjusting ring. When the camera is operated the shutter opens, for a predetermined time, and exposes the film to the light which has entered through the lens.

Fig. 10.5 Camera

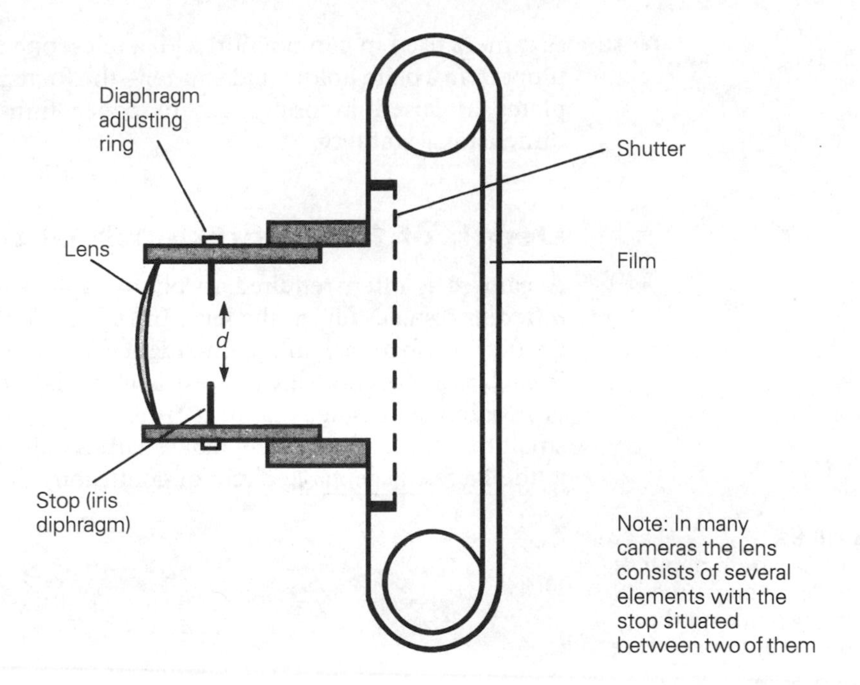

When the object being photographed is moving quickly, the exposure time has to be very short and therefore the diameter of the aperture has to be large in order that the amount of light which falls on the film is sufficient to expose it by the proper amount. When a slow-moving object is photographed, a long exposure can be used and the diameter of the aperture can be reduced. Using a small aperture has two advantages: (i) it reduces spherical aberration because the light which reaches the film has passed only through the central region of the lens, and (ii) it increases the **depth of field**.

f-number

The amount of light that reaches the film is proportional to the area of the aperture, i.e. to d^2. A camera is normally used to photograph objects whose distances from the lens are large compared with its focal length, f. It can be shown that the image of such an object covers an area which is approximately proportional to f^2. It follows that the amount of light per unit area of image is proportional to d^2/f^2 and that the required exposure time is proportional to f^2/d^2. f/d is called the ***f*-number** (or

relative aperture). For a typical camera the available *f*-number settings are 2, 2.8, 4, 5.6, 8, 11, 16, 22. These numbers are such that the square of a number is (approximately) twice that of the number which precedes it. It follows that decreasing the *f*-number by one setting halves the exposure time.

QUESTIONS 10B

1. A camera has an *f*-number of 5.6. When the camera is focused at infinity the centre of the lens is 4.9 cm from the film. What is the diameter of the lens?

2. When a camera is focused on an object 40.0 cm away the lens is 5.7 cm from the film. **(a)** What is the focal length of the lens? **(b)** How far would the lens need to be moved in order to focus the camera on an object at infinity?

Note A camera used in conjunction with a telescope for astronomical purposes is little more than a plate holder and shutter – the focusing is done by the telescope. Glass plates are used in preference to paper film for the sake of durability and dimensional stability.

Depth of field and depth of focus

A camera is often required to photograph, simultaneously, objects that are at different distances from the lens. In Fig. 10.6(a) light from a point object at O is brought to a focus at I on the film. Light from objects at O_1 and O_2 is out of focus and, in each case, covers an area whose diameter is XY and which is known as a **circle of confusion**. The range of object distances for which the circles of confusion are so small that the image is acceptably sharp is called the **depth of field**. Thus if XY is the largest acceptable circle of confusion, the depth of field is O_1O_2.

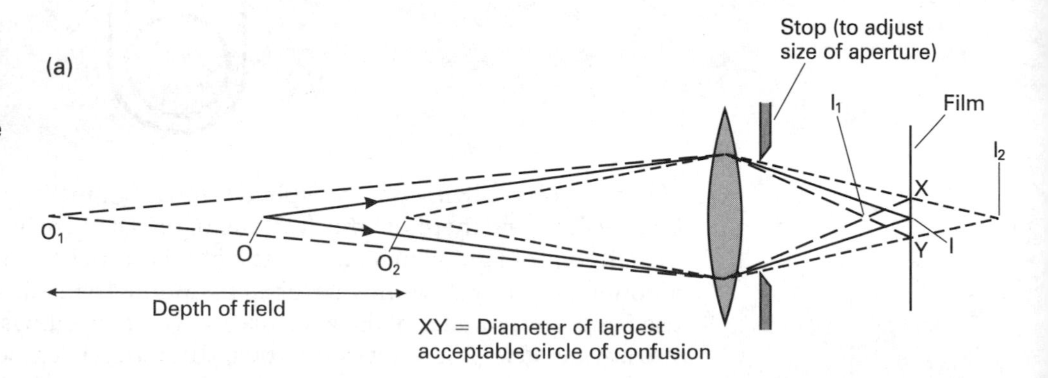

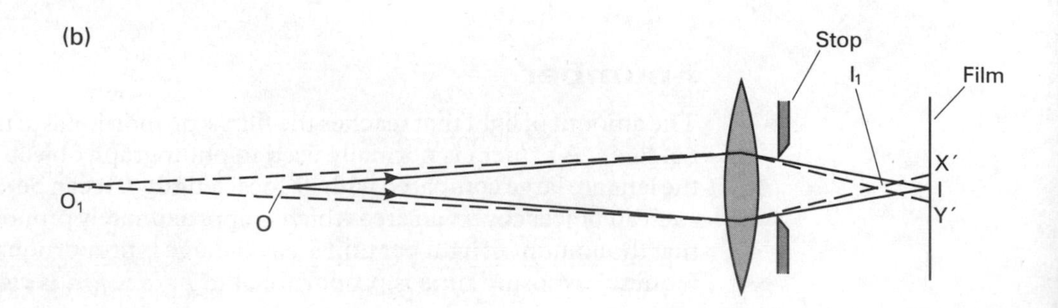

Fig. 10.6 (a) To illustrate depth of field (b) To show the effect of reducing the diameter of the aperture

Decreasing the aperture increases the depth of field because it decreases the diameter of the circle of confusion – X′Y′ in Fig. 10.6(b) is smaller than XY in Fig. 10.6(a).

The range of image distances over which the image of an improperly focused object is acceptably sharp is called the **depth of focus** (Fig. 10.7). It should be clear from Fig. 10.7 that **decreasing the aperture increases the depth of focus** because it decreases the diameter of the circle of confusion.

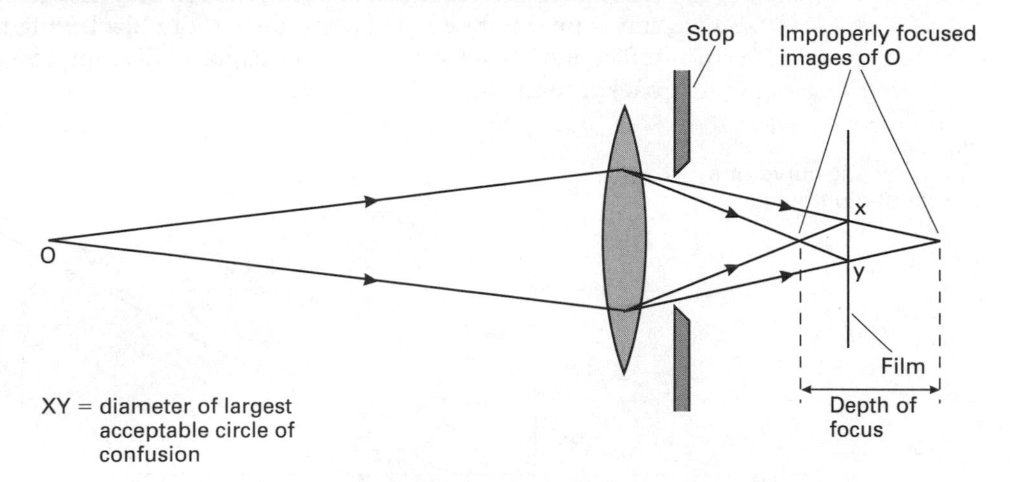

Fig. 10.7 To illustrate depth of focus

10.6 PHOTOGRAPHIC EMULSION

A photographic emulsion consists, essentially, of tiny ($\sim 1\ \mu m$) grains of a silver halide (e.g. silver bromide) in suspension in gelatine. On exposure to light (or UV or X-rays), some of the halide is reduced to silver, producing an (invisible) **latent image**. When a photographic film is developed, a reducing agent such as hydroquinone is used to convert more of the halide to silver in order to create an image that is visible. The silver atoms that were created through exposing the film to light act as catalysts for this process, and therefore it takes place (to any great extent) only in those grains that contributed to the latent image. Thus, the final image is an amplified version of the latent image – typically a billion times as intense. The image is 'fixed' by using sodium thiosulphate to dissolve away all of the silver halide that has not been reduced to silver. This ensures that the image is not degraded by any further exposure to light.

A grain that has been converted to silver looks black and therefore those parts of the film that have been exposed to light appear dark. The greater the exposure, the darker the film. **Image density** is a measure of the extent to which a film has been darkened. It is measured by passing a beam of light through the film. If the intensities of the incident and transmitted beams are respectively I_0 and I, the image density is defined by

$$\text{Image density} = \log_{10}\left(\frac{I_0}{I}\right)$$

Exposure (E) is defined by

$$E = I \times t$$

where I is the intensity (i.e. energy per unit time per unit area) of the light to which the film has been exposed and t is the time for which it has been exposed.

Fig. 10.8 shows how image density depends on exposure for a typical photographic film. The image density reaches **saturation level** at any part of the film where every grain has been turned black. No matter how bright a star actually is, its image will be no more intense than that of a star which is just bright enough to saturate the film. The **gross fog level** is the lowest value of the image density and it is greater than zero even for a film that has been exposed in total darkness. This is because the chemical process of developing a film causes some silver halide grains to blacken even though they have not been exposed to light. If a star is not bright enough to produce more blackening than this purely chemical effect, it does not show up on the exposed film – its image simply merges into the background 'fog'.

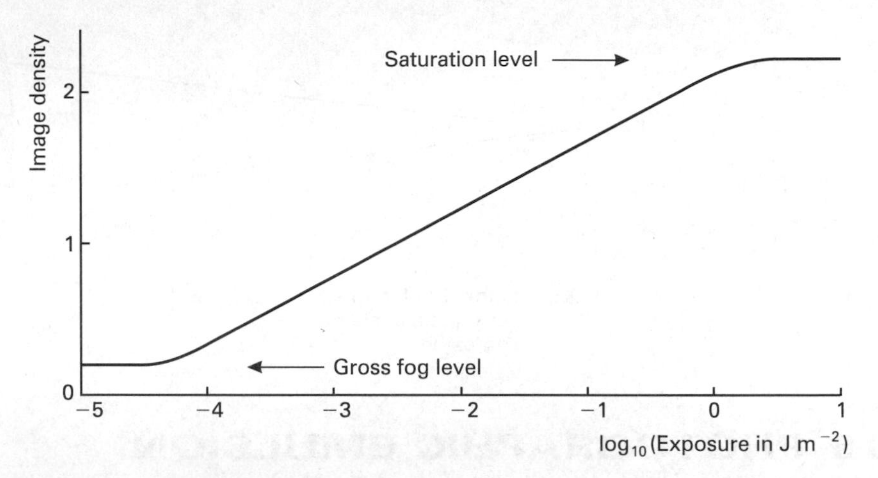

Fig. 10.8 Characteristic curve of a typical photographic emulsion

The quantum efficiency of a typical photographic emulsion is about 4%, i.e. only about 4% of the photons incident on a film cause grain blackening. Of the rest, some may be unproductively captured by halide ions or by grains that have already been triggered, others are reflected from the surface of the film or simply pass straight through it.

Photographic film can register the presence of objects that are far too faint to be seen by the human eye. If a source of light is to be detected by the eye, the intensity it produces at the eye must be above some minimum level. For a photographic emulsion, on the other hand, it is the exposure that must be above a minimum level (the gross fog level). A source of very low intensity will produce an image on photographic film provided the film is exposed to the source for a sufficiently long time; there is no such cumulative effect with the eye.

Another major advantage of photographic film over the human eye is its ability to provide a permanent record.

10.7 THE CHARGE-COUPLED DEVICE (CCD)

This is a silicon wafer whose upper surface is divided into an array of light-sensitive detectors known as picture elements or pixels. One of the latest has over 4 million pixels arranged in a 2048 by 2048 square array measuring 3 cm by 3 cm. When the CCD is exposed to light, charge builds up on the pixels and remains there until it is removed at the completion of the exposure. **The charge that accumulates on any particular pixel is proportional to the total number of photons that has fallen on that pixel.** It follows that the charge at any point on the array is proportional to the brightness at the corresponding point on the image to which the CCD has been exposed. When the exposure is complete, the charge on each pixel is 'read out' into a computer by applying a small voltage to each pixel in turn. This information is then used to create an image on a video monitor.

An astronomical CCD camera. The charge-coupled device (CCD) at its centre has 640 000 light-sensitive elements (pixels) arranged in an array of 800 by 800

CCDs have quantum efficiencies of over 70%. When used in conjunction with the largest telescopes they can detect stars as faint as 29th magnitude, which is several magnitudes better than can be achieved with photographic film. Another advantage of CCDs over photography is that they allow valuable telescope time to be used much more efficiently – what can be achieved in a one minute exposure with a CCD might take as long as two hours photographically. A disadvantage is that, as yet, it is not possible to make CCDs that are larger than about 5 cm by 5 cm and they therefore have a much smaller field of view than is possible with a photographic plate.

CONSOLIDATION

The eye lens provides slight changes in deviation rather than large amounts of it. Most of the deviation occurs at the boundary between the air and the cornea.

The iris controls the intensity of the light reaching the retina by adjusting the size of the **pupil**.

Cones allow us to distinguish different colours and see fine detail but are effective only in bright light.

Rods cannot distinguish different colours and give little perception of detail but allow us to see in dim light.

Depth of field – the range of object distances for which an image is acceptably sharp. Increased by decreasing the size of the aperture.

Depth of focus – the range of image distances over which the image of an object is acceptably sharp. Increased by decreasing the size of the aperture.

$$\text{Image density} = \log_{10}\left(\frac{I_0}{I}\right)$$

Exposure = Intensity × Time for which film is exposed

QUESTIONS ON CHAPTER 10

1. **(a)** Draw a labelled diagram of a human eye. Explain the function of those components which
(i) produce,
(ii) control the intensity of, the image on the retina of an illuminated object.

(b) **(i)** Give the conditions under which rod vision or cone vision in the eye becomes dominant. Hence explain why stars of low apparent brightness all appear to be of the same colour when observed by the unaided eye, but can be seen to be of different colours when seen through a telescope.
(ii) Both the retina of the human eye and a photographic film must receive a minimum of 100 photons on a small area of the sensitive surface in order to record the presence of light. Explain why it is possible to record on a photographic film stars which are too faint to be seen by the naked eye.
[N*]

2. **(a)** There are two types of photoreceptors in the retina of the eye, *cones* and *rods*. State how they differ from each other in their function and response to light.
(b) Sketch a graph to show how the sensitivity of rods varies with wavelength. On the same axes sketch a second graph to show the corresponding variation for cones. Insert approximate numerical values for the wavelength.
(c) Light of wavelength 550 nm falls on the eye from a distant object. If the diameter of the pupil is 3 mm, calculate the smallest angular separation between two objects which the eye can resolve at this distance.
If the retina is 17 mm from the eye lens, estimate how many cones the smallest resolvable patch will cover if each cone has a diameter of the order of 1.1 μm.
[N*, '91]

3. **(a)** In a camera, the lens has a focal length of 50 mm and focused images can be obtained for objects at distances from infinity down to 1.0 m.
Find the minimum and maximum separation between the lens and film needed to provide this facility.
(b) A life-size image of an insect is to be recorded on film.
(i) Find the lens–film separation required using a lens of focal length 50 mm.
(ii) State whether the camera in **(a)** would be suitable for this purpose. Give a reason for your answer. [N, '94]

4. Under certain light conditions a suitable setting for a camera is:

exposure time 1/125 second, aperture $f/5.6$.

If the aperture is changed to $f/16$ what would the new exposure time be in order to achieve the same film image density? What other effect would this change in *f*-number produce?
[L]

5. **(a)** What are the advantages of a camera having:
(i) a lens that can be moved relative to the film,
(ii) a variable aperture?
(b) When a particular camera is focused on infinity the centre of the lens is 56 mm from the film. Calculate the diameter of the aperture when the camera is set to an *f*-number of 2.8.
(c) The correct exposure time with an *f*-number of 2.8 under a particular set of lighting conditions is 1/1000th of a second. What *f*-number would be required for an exposure time of 1/250th of a second under the same lighting conditions? What is the effective diameter of the camera lens at this setting?

6. **(a)** State what is meant by *image density* of an exposed photographic film.
(b) The figure shows the characteristic curve of a panchromatic film in which the image density D is plotted against $\log_{10}(E)$, where E is the exposure of the film. The exposure is a measure of the light energy falling on the film in J m^{-2}.
Explain why there is an upper limit and a lower limit to D.

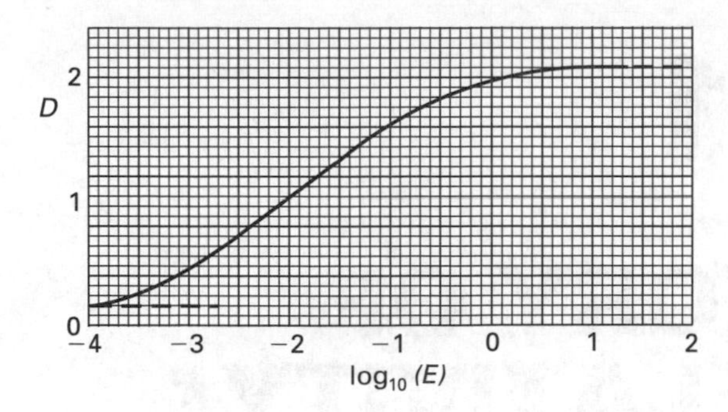

(c) A panchromatic film is exposed to a source of light of wavelength 550 nm. The number of photons arriving at the film is 10^{14} per second and the area of the film is 7.2 cm^2.

Calculate the energy of a single photon arriving at the film and hence, using the data shown in the graph above, show that the film has to be exposed for about 200 seconds before the emulsion becomes saturated.
(The Planck constant, $h = 6.6 \times 10^{-34}$ J s, speed of light $= 3.0 \times 10^8$ m s^{-1}.)
[N*, '94]

7. (a) Describe the mode of operation of a charge-coupled device (CCD) used as a detector in a telescope. Give **two** advantages of using a CCD rather than a photographic plate as a detector.

(b) A CCD is placed at the focal point of a reflecting dish of a telescope. If the dish has a diameter of 2 m and is used to detect radiation of 400 nm wavelength, calculate the smallest angle the dish can resolve. Hence deduce the spacing of the detectors in the CCD if the focal length of the dish is 4 m. [N*, '90]

11

THE THEORY OF SPECIAL RELATIVITY*

11.1 THE ETHER

It used to be believed that there existed throughout the whole of space, and even inside matter itself, a medium known as **the ether**. There has never been any direct evidence for its existence, but it was considered necessary in order to account for the fact that light could travel through a vacuum. This was in the mistaken belief that because sound and all other (non-electromagnetic) wave motions were known to require a medium through which to travel, then so too must light. It was also supposed that **light travels at a fixed speed with respect to the ether**.

Since the ether was supposed to permeate the whole of space, it could be regarded as being a perfect frame of reference relative to which all motion could be measured and which therefore gave meaning to the concept of absolute motion.

11.2 THE MICHELSON–MORLEY EXPERIMENT (1887)

The reader should be familiar with the phenomenon of optical interference before continuing with this section.

(a) Albert Abraham Michelson (1852–1931)
(b) Edward Williams Morley (1838–1923)

(a)

(b)

*For a more complete treatment of the theory of special relativity see R. Muncaster, *Relativity and Quantum Physics* (Stanley Thornes).

If the ether exists, the Earth must be moving through it as it orbits the Sun. A number of experiments have been carried out in attempts to detect this motion. The most famous (and at the time, the most sensitive) was that performed by Michelson and Morley in 1887. If the experiment had had its expected outcome, it would have:

(i) confirmed the existence of the ether, and

(ii) determined the absolute speed of the Earth through space.

Although it failed in both respects, its outcome was described by Bernal as 'the greatest null result in the history of science'.

The principle of the experiment was to compare the speed of light measured in the direction of the Earth's motion with that at right angles to it. The experiment made use of an instrument known as a Michelson interferometer (Fig. 11.1). Light from a monochromatic source at S is partly transmitted and partly reflected on reaching the half-silvered mirror, A. The two beams are then reflected by mirrors M_1 and M_2 respectively, so that they return to A. On reaching A the beams superpose and produce interference fringes which can be seen by an observer at O. (Some light also travels towards the source but this is of no consequence.)

Fig. 11.1
The interferometer used in the Michelson–Morley experiment

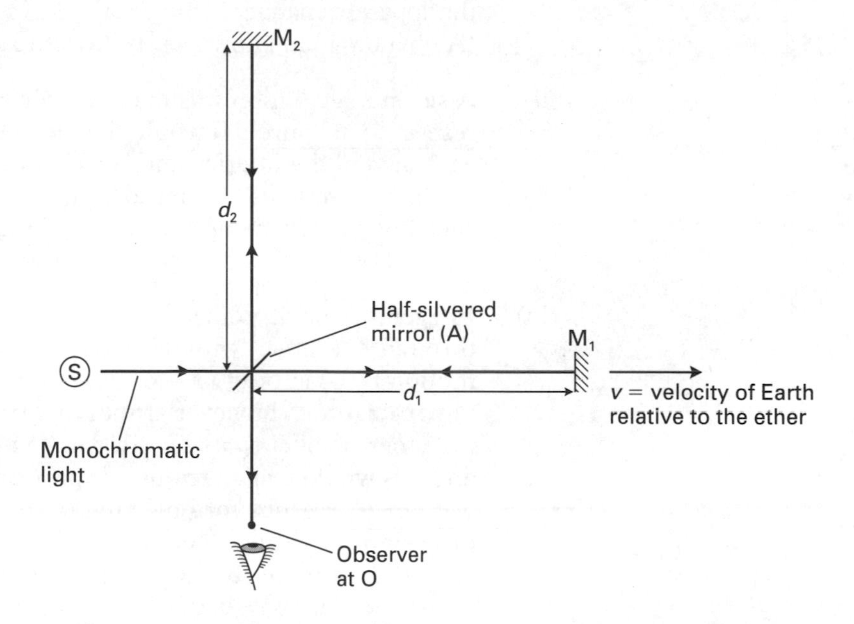

Suppose that the Earth (and therefore the interferometer) is moving to the right. Quite apart from any difference in the distances of M_1 and M_2 from A, because light travels at a fixed speed with respect to the ether, the time taken for light to travel from A to M_1 and back will not be the same as that from A to M_2 and back. The method involved rotating the interferometer through 90° (about a vertical axis) whilst observing the fringes produced at A. This was expected to cause the fringes to shift slightly because one path would now be shorter (in terms of time) and the other would be longer, and this would alter the phase difference between the two beams reaching A.

The pattern was expected to shift by an amount corresponding to just under 2/5 of a wavelength of sodium light. (Half a wavelength would cause each bright fringe to move to the position previously occupied by an adjacent dark fringe etc.) The observed shift corresponded to less than 1/100 of a wavelength, and this could be taken to be zero because there was a possible experimental error of 1/100 of a wavelength. The experiment was repeated on a number of occasions at different times of the year because it was possible that, at the time of the first measurement, the Sun itself was moving through space in just such a way that the Earth was at rest with respect to the ether. The result was always the same – **if the ether existed, the Earth appeared to be at rest relative to it.**

Implications of the result

The Earth moves relative to the stars and to the Sun and to all the planets in the Solar System, and therefore the idea that the Earth really is at rest with respect to the ether can be dismissed immediately for it would make the Earth unique.

Three other ideas were put forward in attempts to explain the result of the Michelson–Morley experiment.

(i) It was suggested by Michelson that the ether in the vicinity of the Earth might be dragged along by the Earth, in which case both arms of the interferometer would be at rest with respect to the ether. This was rejected on the grounds that it was in disagreement with observations concerning the apparent changes in the positions of stars due to the motion of the Earth. (A phenomenon known as **stellar aberration**.)

(ii) A second suggestion was that light always travels at the same speed with respect to its source. It would therefore travel at the same speed relative to each arm of the interferometer. This would make light different from all other wave motions (the speed of sound, for example, is independent of the speed of the source of the sound). In any case, observations of the light emitted by double star systems seemed to rule out this possibility.

(iii) Fitzgerald and Lorentz, independently, proposed that objects moving through the ether with velocity v contract in the direction of their motion by a factor of $(1 - v^2/c^2)^{1/2}$ – the so-called **Fitzgerald–Lorentz contraction**. Whichever arm of the interferometer was moving through the ether parallel to its length would therefore contract by this amount and this would be just enough to produce the observed null result. There were many reasons for objecting to this hypothesis, but none that could prove it to be wrong. Nor could it be proved to be correct. For example, any attempt to verify the contraction by direct measurement would be bound to fail because whatever was used as a 'ruler' would itself contract!

We can draw the following conclusion:

The null result of the Michelson–Morley experiment allows just two possibilities. Either:

(i) the ether does not exist, or

(ii) it does exist but is impossible to detect because the Fitzgerald–Lorentz contraction makes it impossible to detect the motion of any object relative to it.

In either case we have to abandon the idea of absolute motion, and have to accept that all observers will obtain the same value for the speed of light regardless of their velocity.

Theory

Suppose the interferometer (Fig. 11.1) is moving to the right with speed v relative to the ether. If the speed of light relative to the ether is c, then the speed of light relative to the interferometer is $c - v$ from A to M_1 and is $c + v$ from M_1 to A. Therefore

$$\text{Time for path } AM_1A = \frac{d_1}{c-v} + \frac{d_1}{c+v} = \frac{2cd_1}{c^2 - v^2} = \frac{2d_1}{c(1 - v^2/c^2)}$$

$$\approx \frac{2d_1}{c}\left(1 + \frac{v^2}{c^2}\right)^{\star}$$

Fig. 11.2
Diagram for theory of the Michelson–Morley experiment

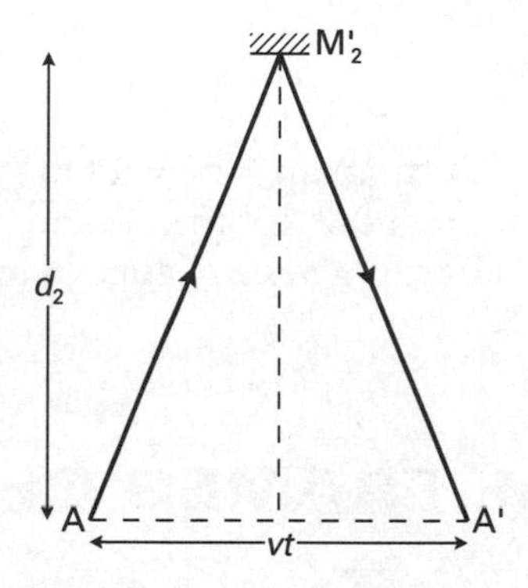

If light travels from A to M_2 and back in time t, the interferometer will move to the right by vt whilst this is happening, and the light path relative to the ether will be as shown in Fig. 11.2, where M_2' is the position of M_2 when the light reaches it and A' is the position of A when the light returns to A. By Pythagoras' theorem

$$(AM_2')^2 = d_2^2 + \left(\frac{vt}{2}\right)^2$$

$$\therefore \quad \left(\frac{ct}{2}\right)^2 = d_2^2 + \left(\frac{vt}{2}\right)^2$$

$$\therefore \quad \frac{t^2}{4}(c^2 - v^2) = d_2^2$$

Therefore

$$\text{Time for path } AM_2A = t = \frac{2d_2}{(c^2 - v^2)^{1/2}} = \frac{2d_2}{c(1 - v^2/c^2)^{1/2}} \approx \frac{2d_2}{c}\left(1 + \frac{v^2}{2c^2}\right)^{\star}$$

$$\therefore \quad \text{Time difference} \approx \frac{2d_1}{c}\left(1 + \frac{v^2}{c^2}\right) - \frac{2d_2}{c}\left(1 + \frac{v^2}{2c^2}\right) = T_0 \text{ say}$$

If the apparatus is rotated through 90 °, then by analogy with what we have just done

$$\text{Time for path } AM_1A \approx \frac{2d_1}{c}\left(1 + \frac{v^2}{2c^2}\right)$$

$$\text{Time for path } AM_2A \approx \frac{2d_2}{c}\left(1 + \frac{v^2}{c^2}\right)$$

*This approximation follows from the binomial expansion.

$\therefore$ Time difference $\approx \dfrac{2d_1}{c}\left(1+\dfrac{v^2}{2c^2}\right) - \dfrac{2d_2}{c^2}\left(1+\dfrac{v^2}{c^2}\right) = T_{90}$ say

$$\text{Change in time difference} \approx T_0 - T_{90}$$

$$\approx \frac{2d_1}{c}\left(\frac{v^2}{2c^2}\right) + \frac{2d_2}{c}\left(\frac{v^2}{2c^2}\right)$$

$$\approx \frac{(d_1 + d_2)\,v^2}{c^3}$$

$\therefore$ Change in number of wavelengths path difference $\approx \dfrac{c}{\lambda} \cdot \dfrac{(d_1 + d_2)\,v^2}{c^3}$

$$\approx \frac{(d_1 + d_2)\,v^2}{\lambda c^2}$$

Putting $d_1 = d_2 = 11\,\text{m}$, $\lambda = 5.9 \times 10^{-7}\,\text{m}$ (the wavelength of sodium light), $c = 3.0 \times 10^8\,\text{m s}^{-1}$ and $v = 3.0 \times 10^4\,\text{m s}^{-1}$ (the mean orbital speed of the Earth) gives a change in path difference of 0.37 wavelengths.

11.3 INERTIAL FRAMES OF REFERENCE

A drinks trolley left unattended in the aisle of an aeroplane which is accelerating down the runway during take off would accelerate towards the back of the plane. This appears to contradict Newton's first law for there is no backward directed force acting on the trolley. Whilst the plane was stationary on the runway prior to take off, the trolley would also be stationary. This, of course, is in total accord with the first law – there is no force on the trolley and it remains at rest. Thus there are some reference frames (e.g. the stationary aeroplane) in which Newton's first law is valid and others (e.g. the accelerating aeroplane) in which it is not. As a second example, consider a particle moving from A to B just above a rotating turntable (Fig. 11.3). An observer at X sees the particle move in a straight line, whereas an observer on the turntable regards the path as being curved. The rotating turntable is therefore another example of a reference frame in which Newton's first law is not obeyed.

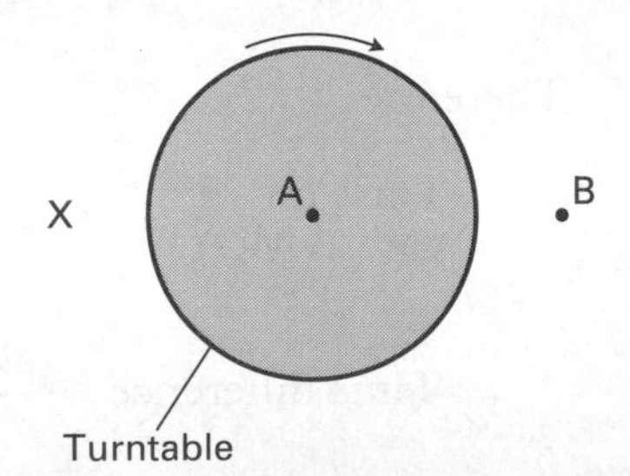

Fig. 11.3
Illustration of a non-inertial reference frame

A reference frame in which Newton's first law is valid is called **an inertial frame of reference.**

It can be shown that if any particular frame is an inertial frame, then any frame which is moving with constant velocity and which is not rotating relative to this frame is also an inertial frame.

Note For most purposes it is a sufficiently good approximation to consider that the Earth is an inertial frame, despite the fact that it is rotating about the Sun.

11.4 THE POSTULATES OF THE THEORY OF SPECIAL RELATIVITY

The theory of special (or restricted) relativity, published by Einstein in 1905, is concerned only with inertial frames of reference. The theory of general relativity (1916) deals with the broader issue of non-uniform relative motion. We shall not be concerned with general relativity.

Albert Einstein (1879–1955)

Special relativity is based on two postulates.

1. The laws of physics have the same form in all inertial frames of reference. (This is known as the **principle of special relativity**.)

2. The speed of light (in vacuum) is the same in all inertial frames; it does not depend on the velocity of either the source or the observer.

The first postulate implies that an experiment performed in one inertial frame will have exactly the same outcome when performed in any other inertial frame. It follows that no one frame can be singled out as being at rest and therefore that **the concept of absolute motion is meaningless, i.e. all motion is relative**. This is entirely consistent with Michelson's and Morley's failure to detect the motion of the Earth through the ether. The second postulate is also in agreement with the outcome of the Michelson–Morley experiment for it means that the speed of light measured in the direction of the Earth's motion is the same as that at 90° to it, and this is what they found.

These simple postulates have far-reaching implications, all of which have been confirmed by experiment. We shall discuss a number of these (time dilation, length contraction, the increase of mass with velocity, the equivalence of mass and energy and the impossibility of accelerating a body beyond the speed of light) in the sections that follow.

Note Time dilation, length contraction and the increase of mass with velocity are noticeable only at speeds which are significant in comparison with the speed of light – see Fig. 11.4.

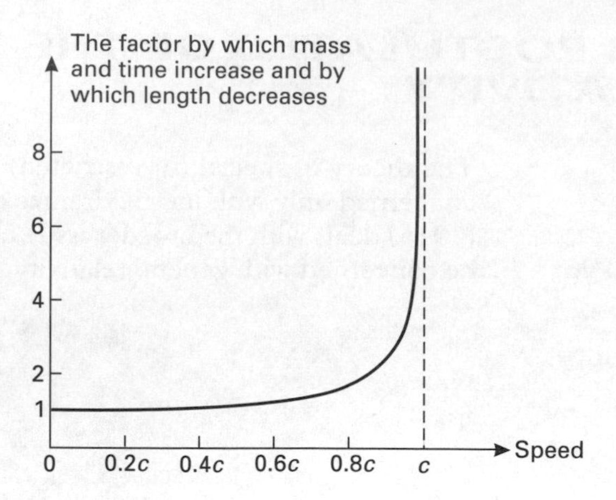

Fig. 11.4
Mass, length and time as a function of speed

11.5 TIME DILATION

According to the theory of special relativity:

> Any observer regards a clock which is moving relative to himself as running slower than a clock which is stationary relative to himself.

We can demonstrate the truth of this by carrying out a **thought experiment**.

Consider two observers, S and T, each of whom has a clock, and suppose that T is on a train travelling to the right relative to S. Suppose also that A and B are two points which are at rest in the train and that a pulse of light travels from A to a mirror at B and then returns to A. T sees the light path as ABA (Fig. 11.5(a)), but S sees it as XYZ (Fig. 11.5(b)) where X and Z are the initial and final positions of A, and Y is the position of the mirror when the pulse reaches it. Since XYZ is greater than ABA, and because light travels at the same speed for all observers (the second postulate), the journey time of the light pulse as measured by S's clock is greater than that measured by T's clock, i.e. S regards T's clock as running more slowly than his own.

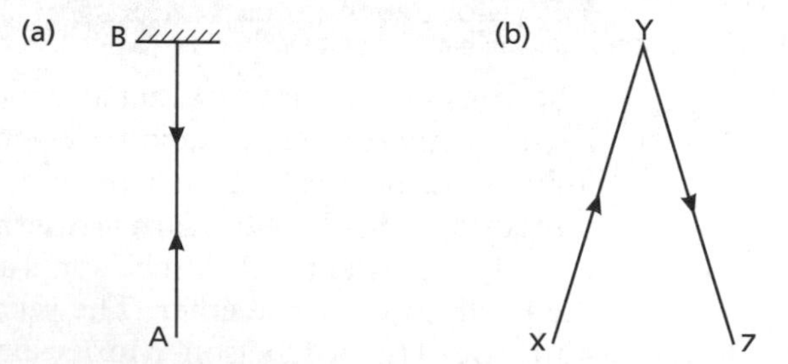

Fig. 11.5
To illustrate time dilation

If the experiment is performed in reverse, i.e. with A and B at rest relative to S, T regards the light pulse as following a longer path than S and therefore regards S's clock as running slow. Thus each observer regards the other's clock as running more slowly than his own.

The reader may have some questions.

Question Which clock is really running slower?

Answer The question is meaningless. We have tried to show that time is relative and this question implies that there is some absolute time because it does not ask from whose point of view. The most we can say is that T's clock is slow as far as S is concerned and that S's clock is the slower one as far as T is concerned.

Question The first experiment shows that S regards T's clock as running slower than his own. Why does this not mean that T regards S's clock as running faster than his?

Answer Imagine that S and T can see each other's clocks. T's clock is at rest with respect to the lamp and mirror and therefore both S and T see the time it registers whilst the light pulse travels from the lamp to the mirror and back. However, S's clock is moving with respect to the lamp and mirror and therefore what T sees S's clock register whilst the pulse travels from the lamp to the mirror and back is not the same as what S sees. The first experiment allows S to compare the two clocks but it does not allow T to.

11.6 EVIDENCE FOR TIME DILATION FROM MUON DECAY

Cosmic ray bombardment of atoms in the upper atmosphere can produce muons – unstable particles which then travel towards the surface of the Earth at speeds close to that of light. In an experiment performed in 1963, Frisch and Smith obtained data on these fast-moving muons that provided convincing evidence of time dilation.

Frisch and Smith compared the rate at which muons moving at $0.994c$ reached a detector on the top of Mt. Washington (USA) with the rate at which muons of the same speed reached a detector 1.9×10^3 m below (Fig. 11.6). As measured by an observer on Earth, this journey takes a time t, where

$$t = \frac{1.9 \times 10^3}{0.994 \times 3.0 \times 10^8} = 6.37\,\mu\text{s}$$

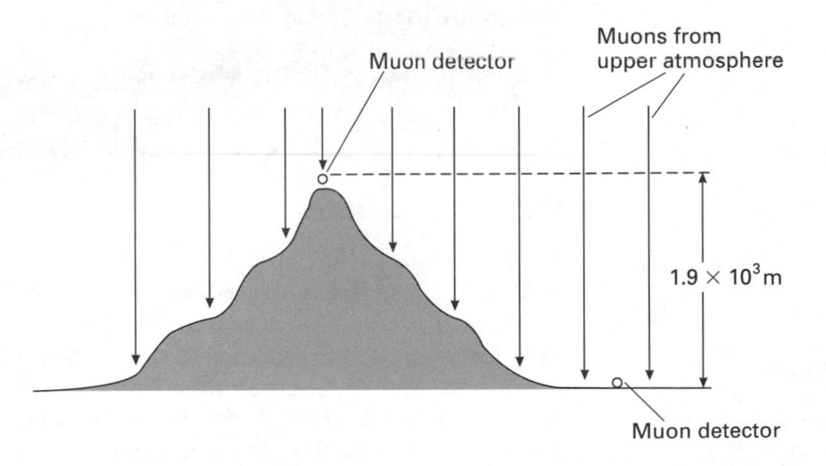

Fig. 11.6 Experiment to confirm time dilation

The half-life of muons as measured by an observer relative to whom they are at rest is $1.53\,\mu$s. Frisch and Smith found that the fraction remaining after travelling 1.9×10^3 m was 0.732, which implies that an observer moving along with the muons would regard the journey as taking a time t_0 given by the exponential law of radioactive decay as

$$0.732 = e^{-(0.6931 t_0/1.53 \times 10^{-6})}$$

i.e. $$t_0 = 0.689\,\mu\text{s}$$

On this basis, then, what an observer who is at rest with respect to the muons regards as 0.689 s is regarded as 6.37 s by an observer who is moving at $0.994c$ with respect to the muons, i.e. the latter regards the former's clock as running slow.

Note It can be shown that

$$t = \frac{t_0}{\sqrt{1 - v^2/c^2}} \qquad [11.1]$$

where v is the relative speed of the two observers and c is the speed of light. When $v = 0.994c$, equation [11.1] gives t/t_0 as 9.1, which compares well with the measured time dilation factor of $6.37/0.689 = 9.2$.

QUESTIONS 11A

1. Use equation [11.1] to explain why time dilation is not noticeable at non-relativistic speeds.

2. A spaceship passes the Earth and flashes a signal lamp for 2 μs as measured by a clock on the spaceship. The duration of the signal as measured on the Earth is 3 μs. At the same time, a lamp on the Earth is switched on for 2 μs as measured by a clock on the Earth. For how long is this lamp on according to a clock in the spaceship?

3. A detector is at such a distance from a source of muons that the rate at which it detects the muons is half that at which they are being produced. An observer who is at rest with respect to the source and the detector measures the journey time from source to detector as 10.5 μs. The half-life of muons is 1.5 μs.
   (a) What is the journey time as measured by an observer moving with the muons?
   (b) At what fraction of the speed of light are the muons travelling?

11.7 LENGTH CONTRACTION

According to the theory of special relativity:

> All observers regard an object which is moving relative to themselves as being shorter in the direction of motion than do observers who are at rest relative to the object.

We can demonstrate the truth of this by a simple thought experiment.

Suppose A and B are two points that are at rest with respect to the Earth, and further suppose that a spaceship travels from A to B at a speed v relative to the Earth in a time t_0 as measured on the spaceship. The pilot of the spaceship therefore regards the distance between A and B as $v \times t_0$. If an observer on the Earth measures the journey time from A to B as t, then he regards the distance between A and B as $v \times t$. However, because of the effect of time dilation, t is greater than t_0 and therefore the Earth observer regards the distance between A and B as being greater than the pilot of the spaceship.

Note It can be shown that

$$L = L_0\sqrt{1 - v^2/c^2}$$

where L_0 is the length of a rod (say) measured by an observer who is at rest relative to the rod, and L is its length as measured by an observer who is moving with a constant velocity v relative to the rod and parallel to its length.

11.8 THE DEPENDENCE OF MASS ON VELOCITY

According to the special theory of relativity the mass of a body is not simply a measure of the amount of material it contains but is a quantity that increases with velocity.

We shall have cause to use the terms 'rest mass' and 'relativistic mass', and we shall define these before going further.

The rest mass (m_0) of a body is its mass as measured by an observer with respect to whom the body is at rest.

The relativistic mass (m) of a body is its mass as measured by an observer with respect to whom it has some velocity other than zero.

It can be shown that

$$m = \frac{m_0}{\sqrt{1 - v^2/c^2}} \qquad [11.2]$$

where m_0 is the rest mass of the body and m is its mass as measured by an observer with respect to whom it has velocity v, i.e. its relativistic mass.

Note The validity of equation [11.2] has been confirmed by experiment (from studies of high-energy electrons, for example) on many occasions.

The impossibility of $v > c$

The denominator of equation [11.2] becomes smaller as v increases, i.e. **relativistic mass increases with velocity** (see Fig. 11.4). Unless $m_0 = 0$, m approaches infinity as v approaches c. This implies that a body with non-zero rest mass* cannot be accelerated up to the speed of light, for if it were to reach the speed of light, it would have infinite mass and this would be absurd. Thus **bodies of non-zero rest mass must always travel at speeds less than that of light**.

The reader may feel that there are two possible objections to this statement.

1. Suppose that two particles are moving in opposite directions, each at a speed of $2c/3$ with respect to some observer. Newtonian mechanics gives their relative speed as $4c/3$, which is clearly greater than the speed of light. According to relativity theory, though, the relative speed v of two particles moving in opposite directions with speeds u_1 and u_2 is given by

$$v = \frac{u_1 + u_2}{1 + \frac{u_1\, u_2}{c^2}}$$

which gives a relative speed of $12c/13$ in this case and this is less than the speed of light. The reader should confirm that even when $u_1 = u_2 = c$, the relative speed is c!

2. Suppose that a powerful laser beam is directed towards the Moon so that a spot of light falls on its surface. If the laser were then turned sideways at the relatively moderate rate of 180 degrees per second, the spot would

* Particles with zero rest mass (e.g. photons) always travel at the speed of light.

move across the Moon at a speed in excess of the speed of light. There is nothing in the theory of relativity to preclude this; neither matter, nor energy, nor information has moved across the Moon.

11.9 THE EQUIVALENCE OF MASS AND ENERGY

According to the theory of special relativity a mass m is equivalent to an amount of energy E, where

$$E = mc^2 \quad [11.3]$$

c being the speed of light ($\approx 3 \times 10^8$ s^{-1}).

It follows that whenever a reaction results in a release of energy there is an associated decrease in mass. For example, when 1 kg of $^{235}_{92}U$ undergoes fission the energy released is approximately 8×10^{13} J, and therefore according to equation [11.3] there is a decrease in mass of $8 \times 10^{13}/(3 \times 10^8)^2 \approx 9 \times 10^{-4}$ kg. This is a significant fraction of the initial mass of $^{235}_{92}U$ and can be measured. Chemical reactions, on the other hand, release relatively small amounts of energy and the associated decrease in mass is too small to be measured. For example, when 1 kg of petrol is burned the energy released is only 5×10^7 J and, by equation [11.3], this corresponds to a decrease in mass of a mere 5.5×10^{-10} kg.

The reader should be left in no doubt that no matter how a change in energy arises there is a change in mass. For example, an increase in temperature is accompanied by an increase in mass, as is an increase in velocity.

CONSOLIDATION

The Ether

The hypothetical medium that was once considered necessary to account for the fact that light can travel through a vacuum.

It was regarded as a perfect frame of reference relative to which all motion could be measured and which therefore gave meaning to the concept of absolute motion.

Light was supposed to travel at a fixed speed (c) with respect to the ether.

The Michelson–Morley Experiment

The aim of the experiment was to measure the speed of the Earth with respect to the ether and therefore

(i) to confirm the existence of the ether, and

(ii) to determine the absolute speed of the Earth.

The null result (i.e. the failure to detect any motion of the Earth through the ether) implies that either

(i) the ether does not exist, or

(ii) it does exist but is impossible to detect.

In either case we have to abandon the idea of absolute motion and have to accept that the speed of light is independent of the velocity of the observer.

The theory of special relativity is based on two postulates:

1. The laws of physics have the same form in all inertial frames of reference.
2. The speed of light (in vacuum) is the same for all inertial observers.

It implies that:

Any observer regards a clock which is moving relative to himself as running slower than a clock which is stationary relative to himself.

All observers regard an object which is moving relative to themselves as being shorter <u>in the direction of motion</u> than do observers who are at rest relative to the object.

The mass of an object as measured by an observer increases as its speed relative to the observer increases.

$$E = mc^2$$

QUESTIONS ON CHAPTER 11

1. **(a)** Why did scientists once believe in the existence of the ether?
(b) Why was the concept eventually abandoned?

2. Give a brief account of the Michelson–Morley experiment. Explain how the outcome of the experiment implied that the concept of absolute motion is meaningless.

3. Explain what is meant by **(a)** time dilation, **(b)** length contraction.

4. **(a)** State the two postulates of Einstein's special theory of relativity.
(b) Muons are created in the upper atmosphere by cosmic rays. They are negatively charged particles with a mass two hundred times that of an electron and a charge of the same size and sign as an electron. They are very short-lived, decaying into an electron and two neutrinos. The graph illustrates the short-lived nature of stationary muons: it shows the number, N, of muons surviving against time, t.

For every 1000 muons detected at a height of 2000 m, 700 are detected at sea level.

(i) Use the graph to estimate how long it would take for 1000 stationary muons to decay to 700.

(ii) How far would a light photon moving through the atmosphere travel in this time?

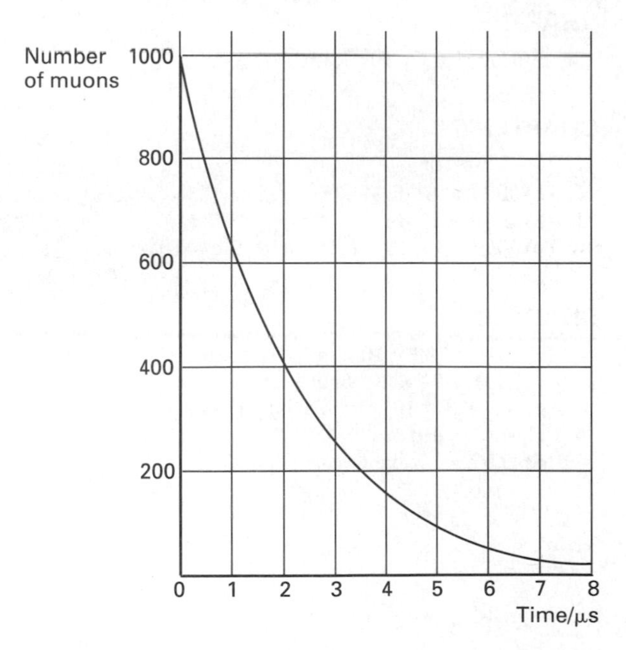

(iii) Muons produced by cosmic radiation travel at a speed of 99.8% of the speed of light. Use the theory of special relativity to explain why such a high percentage of the muons produced by cosmic radiation reach sea level.
(Speed of light $= 3 \times 10^8\ \mathrm{m\,s^{-1}}$)

[N (specimen), '96]

ANSWERS TO END OF CHAPTER QUESTIONS

The Examination Boards accept no responsibility whatsoever for the accuracy or method of working in the answers given. These are the sole responsibility of the author.

CHAPTER 2

3. **(b)** 2.01×10^{30} kg
4. **(b)** 1.1×10^{3} kg m^{-3}
6. **(b)** 5.5×10^{26} kg **(c)** 6.9×10^{-11} N m^{2} kg^{-2}

CHAPTER 3

1. **(b)** **(i)** 7.25×10^{3} K **(ii)** 9.37×10^{26} W
2. **(a)** 2.9×10^{2} K **(b)** 3.0×10^{11} m
8. **(c)** 1.91
9. **(a)** 2.51 **(b)** 5
10. **(a)** 3.91×10^{26} W

CHAPTER 4

3. **(c)** **(ii)** 3.0×10^{2} parsecs

CHAPTER 5

3. 2.96 km, 1.83×10^{19} kg m^{-3}

CHAPTER 6

1. **(c)** 3.7×10^{4} s
2. **(b)** 1.71×10^{4} m s^{-1}
3. **(b)** 1.10×10^{4} m s^{-1}
4. **(b)** **(ii)** 2.3×10^{5} m s^{-1} **(iii)** 9.0×10^{9} m

CHAPTER 7

2. **(a)** **(ii)** 4.733×10^{-7} m **(iii)** 6.09×10^{7} m s^{-1}
(iv) 1.99×10^{9} light-years
(c) **(i)** 9.3×10^{25} m (radius) **(ii)** 3.09×10^{17} s
6. **(c)** 6.09×10^{7} m s^{-1}
(d) 3.97×10^{9} light-years

CHAPTER 8

1. **(a)** 550 mm **(d)** 10 **(e)** 55 mm from eyepiece lens
3. 400 mm (objective), 50 mm (eyepiece)
4. **(b)** 60 mm
5. 4.0 cm
6. 112.4 cm, 15.6
7. **(a)** 160 **(b)** 25.2 mm **(c)** 0.938 mm
8. **(c)** **(i)** 1.84 cm
9. 0.18(2) cm; 20 cm from second lens on the same side as the first lens; 0.91 cm
10. **(a)** 2.1 cm **(b)** 112.5 cm
11. **(b)** **(ii)** 104.2 cm, 24 **(iii)** 5.3 cm behind eyepiece
(c) **(i)** 106.7 cm
13. **(a)** 0.029 cm **(b)** 0.55 cm **(c)** 0.11 rad

CHAPTER 9

2. **(b)** **(i)** 1.1×10^{30} W **(ii)** 2.2×10^{6} light-years
3. **(a)** **(i)** 21.0 cm
4. **(c)** **(i)** 21.1 cm
6. **(c)** 1.50×10^{11} m (Earth), 1.08×10^{11} m (Venus)
8. 2.10×10^{7} s

CHAPTER 10

2. **(c)** 2.2×10^{-4} rad, 12
3. **(a)** 50 mm, 52.6 mm **(b)** **(i)** 100 mm
4. 0.065 s
5. **(b)** 20 mm **(c)** 5.6, 10 mm
6. **(c)** 2.6×10^{-19} J
7. **(b)** 2.44×10^{-7} rad, 0.98 μm

ANSWERS TO QUESTIONS 2A–11A

QUESTIONS 2A

2. 1.9 years

QUESTIONS 2B

1. 2.7×10^{-6} N
2. 6.0×10^{24} kg
3. 1.90×10^{27} kg

QUESTIONS 2C

1. **(a)** 2.7×10^{30} kg **(b)** 2.2×10^{30} kg

QUESTIONS 3A

1. 5.78×10^{3} K
2. 1.39×10^{6} km

QUESTIONS 3B

1. 1.5
2. **(a)** 10 to 1 **(b)** 100 to 1
3. 23 to 1
4. 3.13
5. **(a)** 3.87×10^{26} W **(b)** 5.79×10^{3} K

QUESTIONS 3C

1. 0.60
2. 2.4×10^{2} parsecs
3. **(a)** 5 **(b)** 3
4. **(b)** 10

QUESTIONS 3D

1. 4×10^{2}
2. 1×10^{-2}

QUESTIONS 4A

1. **(a)** 3.4 parsecs **(b)** 11 light-years

QUESTIONS 4B

1. 5.5×10^{5} parsecs

QUESTIONS 5A

1. −2
2. 1.85×10^{-3} kg m^{-3}

QUESTIONS 5B

1. 8×10^{7} years

QUESTIONS 6A

1. **(b)** 4.26×10^{6} m s^{-1} **(c)** 4.9303×10^{-7} m

QUESTIONS 6B

1. **(a)** 3.02×10^{5} m s^{-1} **(b)** 8.30×10^{11} m
2. 4.57×10^{5} m s^{-1}

QUESTIONS 7A

1. **(a)** 3.1×10^{25} m **(b)** 3.3×10^{9} light-years
2. 80 Mpc
3. 1.3×10^{10} years

QUESTIONS 7B

1. 5.3×10^{39} W

QUESTIONS 8A

1. 72 cm
2. **(a)** 85.0 cm **(b)** 16 **(c)** 0.40 cm

QUESTIONS 8B

1. **(a)** 1.3×10^{-7} radians **(b)** 0.027 arc seconds
2. 74 km

QUESTIONS 9A

1. 8.9×10^{4} s

QUESTIONS 10B

1. 0.88 cm
2. **(a)** 5.0 cm **(b)** 0.7 cm

QUESTIONS 11A

2. 3 μs
3. **(a)** 1.5 μs **(b)** 0.99

INDEX